Studer
Study Guide
Cheryl Cantwell

Prealgebra
THIRD EDITION

K. Elayn Martin-Gay

Prentice
Hall

Upper Saddle River, NJ 07458

Executive Editor: Karin E. Wagner
Project Manager: Mary Beckwith
Special Projects Manager: Barbara A. Murray
Production Editor: Wendy A. Perez
Supplement Cover Manager: Paul Gourhan
Supplement Cover Designer: PM Workshop Inc.
Manufacturing Buyer: Alan Fischer

Printed in the United States of America

10 9 8 7 6 5 4 3 2 1

ISBN 0-13-026462-8

Prentice-Hall International (UK) Limited, London
Prentice-Hall of Australia Pty. Limited, Sydney
Prentice-Hall Canada, Inc., Toronto
Prentice-Hall Hispanoamericana, S.A., Mexico
Prentice-Hall of India Private Limited, New Delhi
Pearson Education Asia Pte. Ltd., Singapore
Prentice-Hall of Japan, Inc., Tokyo
Editora Prentice-Hall do Brazil, Ltda., Rio de Janeiro

TABLE OF CONTENTS

CHAPTER 1 WHOLE NUMBERS AND INTRODUCTION TO ALGEBRA

1.1 Place Value and Names for Numbers ... 1
1.2 Adding Whole Numbers and Perimeter ... 2
1.3 Subtracting Whole Numbers ... 3
1.4 Rounding and Estimating ... 5
1.5 Multiplying Whole Numbers and Area ... 6
1.6 Dividing Whole Numbers ... 8
1.7 Exponents and Order of Operations ... 9
1.8 Introduction to Variables and Algebraic Expressions ... 10
Practice Test ... 12

CHAPTER 2 INTEGERS

2.1 Introduction to Integers ... 14
2.2 Adding Integers ... 15
2.3 Subtracting Integers ... 16
2.4 Multiplying and Dividing Integers ... 17
2.5 Order of Operations ... 18
Practice Test ... 19

CHAPTER 3 SOLVING EQUATIONS AND PROBLEM SOLVING

3.1 Simplifying Algebraic Expressions ... 20
3.2 Solving Equations: The Addition Property ... 21
3.3 Solving Equations: The Multiplication Property ... 23
3.4 Solving Linear Equations in One Variable ... 24
3.5 Linear Equations in One Variable and Problem Solving ... 26
Practice Test ... 28

CHAPTER 4 FRACTIONS

4.1 Introduction to Fractions and Equivalent Fractions ... 30
4.2 Factors and Simplest Form ... 31
4.3 Multiplying and Dividing Fractions ... 33
4.4 Adding and Subtracting Fractions and Least Common Denominator ... 34
4.5 Adding and Subtracting Unlike Fractions ... 36
4.6 Complex Fractions and Review of Order of Operations ... 38
4.7 Solving Equations Containing Fractions ... 40
4.8 Operations on Mixed Numbers ... 42
Practice Test ... 44

CHAPTER 5 DECIMALS
    5.1 Introduction to Decimals   46
    5.2 Adding and Subtracting Decimals   47
    5.3 Multiplying Decimals and Circumference of a Circle   48
    5.4 Dividing Decimals   50
    5.5 Estimating and Order of Operations   51
    5.6 Fractions and Decimals   53
    5.7 Equations Containing Decimals   54
    5.8 Square Roots and the Pythagorean Theorem   55
    Practice Test   56

CHAPTER 6 RATIO AND PROPORTION
    6.1 Ratios   58
    6.2 Rates   59
    6.3 Proportions   60
    6.4 Proportions and Problem Solving   61
    6.5 Similar Triangles and Problem Solving   63
    Practice Test   66

CHAPTER 7 PERCENT
    7.1 Percents, Decimals and Fractions   68
    7.2 Solving Percent Problems with Equations   69
    7.3 Solving Percent Problems with Proportions   70
    7.4 Applications of Percent   72
    7.5 Percent and Problem Solving: Sales Tax, Commission, and Discount   74
    7.6 Percent and Problem Solving: Interest   75
    Practice Test   77

CHAPTER 8 GRAPHING AND INTRODUCTION TO STATISTICS
    8.1 Reading Circle Graphs   79
    8.2 Reading Pictographs, Bar Graphs, and Line Graphs   81
    8.3 The Rectangular Coordinate System   84
    8.4 Graphing Linear Equations   86
    8.5 Mean, Median, and Mode   87
    8.6 Counting and Introduction to Probability   88
    Practice Test   89

CHAPTER 9 GEOMETRY AND MEASUREMENT

    9.1 Lines and Angles   92

    9.2 Linear Measurement   93

    9.3 Perimeter   94

    9.4 Area and Volume   96

    9.5 Weight and Mass   98

    9.6 Capacity   99

    9.7 Temperature   100

    Practice Test   102

CHAPTER 10 POLYNOMIALS

    10.1 Adding and Subtracting Polynomials   104

    10.2 Multiplication Properties of Exponents   105

    10.3 Multiplying Polynomials   106

    10.4 Introduction to Factoring Polynomials   107

    Practice Test   108

PRACTICE FINAL EXAMINATIONS   110

SOLUTIONS   118

Study for Success

Congratulations! You have made a responsible decision in owning this Study Guide that supplements your textbook <u>Prealgebra, 3e.</u> Elayn Martin-Gay. This Study Guide provides you with additional resources to help you be successful in your mathematics course.

In the first part of the Study Guide there are examples, exercises, and hints and warnings to supplement each section of the textbook. Also there is a practice test to accompany each chapter and two practice final examinations. All of the examples, exercises, and tests are similar to those in the textbook.

At the beginning of each section the examples with their solutions are presented. These are followed by an exercise set of 15-30 problems. Next a practice test is presented. At the end of each chapter Helpful Hints and Insights are given. At the end of all of the chapters two 100 question practice final examinations are featured.

In the second part of the Study Guide complete step-by-step solutions are worked out for all exercises, tests, and examinations. The methods of solving the problems are just like those presented in the textbook.

Consider the following tips for becoming more successful in your mathematics course.

- Attend class everyday. If you have to miss a class, borrow the notes from another student and make a copy. Call and get the homework assignment so that you do not get behind. Watch the video that accompanies the chapter you missed.

- In the classroom be sure you have a good view of the board and that you can hear.

- Come to class prepared. Have a notebook just for this class, preferably consisting of a section for notes and a section for homework. Bring writing utensils.

- Since a lecture can move rapidly you are forced to take notes quickly. Set up your own abbreviation system. Take notes as thoroughly as possible and after class you can re-write your notes more legibly and fill in the gaps.

- Be sure you know what is assigned for homework. Do your homework as soon after lecture as possible. This way the information is still fresh in your mind.

- When reading your textbook, don't just read the examples. The entire section is important in your understanding of the material.

- As you do homework problems and read the textbook, write down your questions. Don't rely on your memory, because when you get to class you will most likely have forgotten some of your questions.

- Don't be afraid to ask questions. If something is not clear ask for further explanation. Usually there are other students with the same questions.

- When doing homework or taking a test give yourself plenty of paper. Show every step of a problem in an organized manner. This will help reduce the number of careless errors.

- It is important to study your mathematics as many days as you can possibly fit into your schedule. Even if it is just for 10 or 15 minutes, this is beneficial. Do not wait until just before a test to do a crash study session. The material needs to be understood and learned, not memorized. Most likely you will have another mathematics course after this one. You are gaining tools in this class that you will use in the next class. **Retention is the key word.**

- Of course you can't remember everything, that's why it's important to have good notes. Hopefully in the next class even if you can't remember exactly how to do a problem, you will remember that you have done it before and you can find it in your notes. Then you refresh your memory more quickly.

- When doing your homework try each type of problem that was assigned. It is easy to sit in a lecture and a problem appear very simple as the instructor does the work. However, when you are on your own it may all of a sudden become more difficult.

- Practice is what improves your skills in every area including mathematics. Practice is what will make you become comfortable with problems. If there are certain assigned problems that you are having difficulty with, do more of this type until you can do them easily. Remember this Study Guide provides you with lots of additional problems. Once you have done them you can then check them with the completely worked out solutions that are also provided.

- Make use of resources that accompany the textbook. Watch the videos and use the CD tutorial for more practice.

- Have assignments in on time. Do not lose points needlessly.

- Take advantage of all extra-credit opportunities if they exist.

- Once more let me stress the importance of preparing daily for a test not just the day before a test.

- Many times students can do their homework assignments, but then "freeze-up" on a test. Their minds go blank, they panic and hence they do not do well on their test. If this happens to you, it can be overcome by practicing taking tests. When studying for a test actually make yourself a test. Then find a quiet spot and pretend you are in class. Take this test just as if you were in class. The more you practice taking tests, the more comfortable you will be in class during the actual event. Don't forget the Practice Tests provided in this Study Guide. When you use them you will not only be able to check your answers, but also your work since complete solutions are provided.

- When your instructor hands back a graded test, be sure to make corrections to any problems you missed. It is important to clear up these mistakes now because you will generally be using these concepts again. Also when exam time rolls around, your old tests are a great study source. Remember there are two Practice Final Examinations along with complete solutions provided in this Study Guide. Use it to prepare for your exam.

- If you have taken advantage of the previous tips, but you are still having a lot of difficulty, look for other sources of help. See your instructor during office hours, work with a fellow class-mate, form a study group, go to the math lab if one exists at your school, or get a tutor. Don't let things get out of hand, get help early. Be a responsible student.

Good luck in your mathematics course! Hopefully this Study Guide and these helpful hints will contribute to your success in this course. However, remember it is up to you to be a responsible student and take advantage of all resources provided and take all of the steps necessary to being successful!

Cheryl V. Cantwell
Seminole Communtiy College

1.1 PLACE VALUE AND NAMES FOR NUMBERS

Example 1: Write 346,125 in words.

Solution:

three hundred forty-six   **thousand,**   one hundred twenty-five

     number in   name of   number in
      period   period   period

Example 2: Write sixteen thousand, nine-hundred eighty-seven in standard form.

Solution: 16,987

Example 3: Write 1,203,498 in expanded form.

Solution: 1,000,000 + 200,000 + 3000 + 400 + 90 + 8

Example 4: Insert < or > to make a true statement.
48   95

Solution: 48 < 95

1.1 EXERCISES

Determine the place value of the digit 4 in each whole number.

1. 2418
2. 45,691
3. 435,019
4. 47,912,030

Write each whole number in words.

5. 7391
6. 20,068
7. 127,455
8. 5040

Write each whole number in standard form.

9. Five thousand, thirty-one

10. Eight million, two hundred forty-one thousand, seven

11. Twenty-eight thousand, four

12. Two billion, five thousand, nine

Write each whole number in expanded form.

13. 531
14. 7809
15. 61,738,425
16. 40,050,061

Insert < or > to make a true statement.

17. 9   21
18. 10   2
19. 31   29
20. 51   61

The table below shows the 5 longest rivers in the world. Use this table to answer Exercises 21 and 22.

River	Miles
Chang jiang – Yangtze (China)	3964
Amazon (Brazil)	4000
Tenisei–Angara (Russia)	3442
Mississippi – Missouri (U.S.)	3740
Nile (Egypt)	4145

21. Write the length of the Nile River in words.

22. Write the length of the Chang jiang – Yangtze River in expanded form.

1.2 ADDING WHOLE NUMBERS AND PERIMETER

Example 1: Add: 241,836 + 54,791

Solution:

$$\begin{array}{r} {}^{1}{}^{1} \\ 241,836 \\ +54,791 \\ \hline 296,627 \end{array}$$

Example 2: Rewrite the sum 92 + 81 using the commutative property of addition.

Solution: 92 + 81 = 81 + 92

Example 3: Find the sum : 41 + 298 + 1350 + 165.

Solution:

$$\begin{array}{r} {}^{1} \\ {}_{2}41 \\ 298 \\ 1350 \\ +165 \\ \hline 1854 \end{array}$$

Example 4: Find the perimeter of the given triangle.

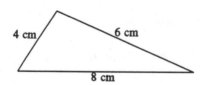

Solution: Add the lengths of the sides:

4 cm + 6 cm + 8 cm = 18 cm
The perimeter is 18 cm.

Example 5: Carl Wilson works at Subs To Go. His monthly salary of $1740 is increased by a raise of $235. What is his new salary?

Solution:

old salary	→	$1740
+ increase	→	235
new salary		$1975

His new salary is $1975.

1.2 EXERCISES

Add.

1. 19 + 36

2. 81 + 75

3. 123 + 985

4.   27
     + 47

5.   50
     + 49

6.   378
     + 124

7. 3246 + 289

8. 5177 + 4960

9. 50 + 47 + 68

10. 18 + 134 + 298 + 1357

11. 19 + 31 + 18 + 32 + 61

12.   1972
      + 8314

13.   12,388
      + 9,615

14.   22
        378
      + 95,416

15.   4000
       200
      + 6091

16.   321,475
       109,316
       25,492
      + 87,500

Find the perimeter.

17.   3 feet [rectangle] 7 feet

18.

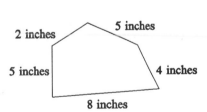

Solve.

19. A suit sells for $189 plus a markup of $36. Find the total price of the suit.

20. Andrea Roberts and Kelsey Ohleger are going to pool their money to buy a new video game costing $62. Andrea has $47 and Kelsey has $18. Determine whether they have enough money to buy the game.

1.3 SUBTRACTING WHOLE NUMBERS

Example 1: Subtract, and then check the answer by adding.

18 − 7

Solution: 18 − 7 = 11 because 11 + 7 = 18

3

Example 2: Find 9645 - 7213, and then check by adding.

Solution:
$$\begin{array}{r} 9645 \\ -\ 7213 \\ \hline 2432 \end{array}$$

Check:
$$\begin{array}{r} 2432 \\ +\ 7213 \\ \hline 9645 \end{array}$$

Example 3: Subtract 805 - 126. Then check by adding.

Solution: Check:

$$\begin{array}{r} ^{\;\;\;\;9} \\ ^{7\ \ 10\ 15} \\ 8\ 0\ 5 \\ -1\ 2\ 6 \\ \hline 6\ 7\ 9 \end{array}$$

$$\begin{array}{r} ^{1\ \ 1} \\ 6\ 7\ 9 \\ +1\ 2\ 6 \\ \hline 8\ 0\ 5 \end{array}$$

Example 4: Tai is reading a 472-page book. If she has just finished reading page 138, how many more pages must she read to finish the book.

Solution:
        total number of pages
       −number of pages read
        number of pages remaining

$$\begin{array}{r} ^{\;\;\;6\ 12} \\ 4\ 7\ 2 \\ -1\ 3\ 8 \\ \hline 3\ 3\ 4 \end{array}$$

She has 334 pages left to read.

1.3 EXERCISES

Subtract and check by adding.

1. $\begin{array}{r} 59 \\ -34 \\ \hline \end{array}$

2. $\begin{array}{r} 65 \\ -25 \\ \hline \end{array}$

3. $\begin{array}{r} 873 \\ -642 \\ \hline \end{array}$

4. $\begin{array}{r} 613 \\ -292 \\ \hline \end{array}$

5. $\begin{array}{r} 81 \\ -69 \\ \hline \end{array}$

6. $\begin{array}{r} 53 \\ -47 \\ \hline \end{array}$

7. $\begin{array}{r} 299 \\ -187 \\ \hline \end{array}$

8. $\begin{array}{r} 500 \\ -199 \\ \hline \end{array}$

9. $\begin{array}{r} 401 \\ -58 \\ \hline \end{array}$

10. 51 - 37

11. 28 - 16

12. 121 - 93

13. 517 - 468

14. 207 - 99

15. 2317 - 1809

16. Subtract 13 from 35.

17. Find the difference of 31 and 16.

18. When Wendy Chisholm began a trip, the odometer read 61,894. When the trip was over the odometer read 63,271. How many miles did she drive on the trip?

19. A television that normally sells for $799 is discounted by $87 in a sale. What is the sale price?

20. On one day in June the temperature in Manassas, Virginia, dropped 13 degrees from 5 p.m. to 8 p.m. If the temperature at 5 p.m. was 81 degrees, what was the temperature at 8 p.m.?

1.4 ROUNDING AND ESTIMATING

Example 1: Round 49,873 to the nearest hundred.

Solution:

49,8<u>7</u>3   The digit to the right of the hundreds place is
↑   the tens place, which is underlined.
hundreds place

49,8<u>7</u>3   Since the underlined digit is 5 or greater, add 1 to the 8 in
↑~~   the hundreds place and replace each digit to the right by 0.
Add Replace
1   with zeros

49,873 rounded to the nearest hundred is 49,900.

Example 2: Round 324,597 to the nearest ten-thousand.

Solution:

Ten-thousands place ——— 4 is less than 5
↓ ↓
324,597
↑ ~~~
Do not ↑ Replace with zeros
add 1

The number 324,597 rounded to the nearest ten-thousand is 320,000.

Example 3: Round each addend to the nearest hundred to find an estimated sum.

```
   365
   717
  2068
+  129
```

Solution:

```
   365   rounds to    400
   717   rounds to    700
  2068   rounds to   2100
+  129   rounds to  + 100
                     3300
```

The estimated sum is 3300. (The exact sum is 3279.)

1.4 EXERCISES

Round each whole number to the given place.

1. 591 to the nearest hundred

2. 372 to the nearest ten

3. 687 to the nearest hundred

4. 3918 to the nearest thousand

5. 8913 to the nearest thousand

6. 1974 to the nearest hundred

7. 16,399 to the nearest hundred

8. 21,589 to the nearest thousand

9. 73,694 to the nearest ten–thousand

10. 49,999 to the nearest ten

Estimate the sum or difference by rounding each number to the nearest ten.

11.   38
      47
      21
   + 15

12.   86
      74
      63
   + 99

13.   731
   – 187

14.   918
     461
  + 379

15.   604
   – 399

16.   2673
  – 1892

Round each given number to the indicated place and solve.

17. Willy, Tom, and Mandy have $436, $1234, and $765, respectively. Estimate to the nearest $100 the total amount of money they have.

18. Enrollment figures at the local unversity showed an increase from 52,821 credit hours in 1998 to 55,674 credit hours in 1999. Estimate the increase to the nearest thousand credit hours.

1.5 MULTIPLYING WHOLE NUMBERS AND AREA

Example 1: Multiply: 32
           × 4

Solution: 32
     × 4
    128

Example 2: Rewrite $7(3 + 6)$ using the distributive property.

Solution: $7(3 + 6) = 7 \cdot 3 + 7 \cdot 6$

Example 3: Multiply: 125×72

Solution:

$$
\begin{array}{rl}
125 & \\
\underline{\times\ 72} & \\
250 & \quad 2(125) \\
\underline{8750} & \quad 70(125) \\
9000 & \quad \text{Add.}
\end{array}
$$

Example 4: Sam and Lynn DeVaughn plan to take their four children to the movies. The ticket price for each child is $4 and for each adult, $7. How much money is needed for their admission?

Solution:

price of 2 adults	$ 14	$(2 \cdot 7)$
+ price of 4 children	+ 16	$(4 \cdot 4)$
total cost	$ 30	

The total cost is $30.

1.5 EXERCISES

Use the distributive property to rewrite each expression.

1. $5(2 + 7)$

2. $8(1 + 4)$

3. $12(20 + 3)$

Multiply.

4. $\begin{array}{r} 39 \\ \underline{\times\ 3} \end{array}$

5. $\begin{array}{r} 21 \\ \underline{\times\ 7} \end{array}$

6. $\begin{array}{r} 236 \\ \underline{\times\ 7} \end{array}$

7. $\begin{array}{r} 243 \\ \underline{\times\ 21} \end{array}$

8. $\begin{array}{r} 798 \\ \underline{\times\ 74} \end{array}$

9. $\begin{array}{r} 2134 \\ \underline{\times\ 62} \end{array}$

10. $\begin{array}{r} 869 \\ \underline{\times\ 20} \end{array}$

11. $\begin{array}{r} 124 \\ \underline{\times\ 231} \end{array}$

12. $\begin{array}{r} 2645 \\ \underline{\times\ 237} \end{array}$

13. $(80)(70)$

14. $(19)(1)(30)$

15. $(297)(31)(0)$

Estimate the products by rounding each factor to the nearest hundred.

16. 483×291

17. 213×365

Find the area of the given rectangle.

18.

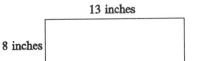

13 inches

8 inches

19.

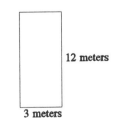

12 meters

3 meters

20. A computer printer can print 65 characters per second in letter-quality mode. Find how many characters it can print in 38 seconds.

1.6 DIVIDING WHOLE NUMBERS

Example 1: Find the quotient and then check the answer by multiplying.

$$56 \div 8$$

Solution: $56 \div 8 = 7$ because $7 \cdot 8 = 56$.

Example 2: Divide: $2572 \div 4$. Check the answer by multiplying.

Solution:

```
      643
   4)2572
     24        6(4) = 24
     17        25 - 24 = 1, bring down, 7
     16        4(4) = 16
     12        17 - 16 = 1, bring down 2
     12        4(3) = 12
      0        12 - 12 = 0
```

Check:
```
   643
 ×   4
  2572
```

Example 3: Find $2851 \div 14$.

Solution:

```
     203  R 9
  14)2851
     28        2(14) = 28
     05        Subtract and bring down the 5.
      0        0(14) = 0
     51        Subtract and bring down the 1.
     42        3(14) = 42
      9        Subtract.
```

Example 4: How many boxes are needed to ship 62 pairs of boots to Colorado if 8 pairs of boots will fit in each shipping box?

Solution: (number of boxes) = (total pairs of boots) ÷ (How many pairs in a box)

number of boxes = 62 ÷ 8

```
      7
   8)62
     56
      6
```

Seven full boxes with 6 pairs of boots left over, so 8 boxes will be needed.

1.6 EXERCISES

Divide, and then check by multiplying.

1. $91 \div 7$
2. $95 \div 5$
3. $152 \div 4$

4. $234 \div 5$
5. $329 \div 3$
6. $987 \div 21$

7. $321 \div 4$
8. $601 \div 5$
9. $5505 \div 11$

10. $6907 \div 17$
11. $3638 \div 36$
12. $2056 \div 54$

13. $\dfrac{6528}{192}$
14. $\dfrac{13167}{231}$
15. $\dfrac{13324}{120}$

Find the average of each list of numbers.

16. 91, 86, 74, 99, 115

17. 18, 23, 14, 26, 28, 31, 7

Solve.

18. Thirteen people pooled their money and bought lottery tickets. One ticket won a prize of $4,485,000. Find how many dollars each person receives.

19. Light poles along a highway are placed 483 feet apart. Find how many poles there are along 2 miles of highway. (A mile is 5280 feet.)

20. Find how many yards there are in 2 miles. (A mile is 5280 feet; a yard is 3 feet.)

1.7 EXPONENTS AND ORDER OF OPERATIONS

Example 1: Write using exponential notation.

     $3 \cdot 3 \cdot 3 \cdot 3 \cdot 3$

Solution: $3 \cdot 3 \cdot 3 \cdot 3 \cdot 3 = 3^5$ The base 3 is multiplied times itself 5 times.

Example 2: Evaluate 4^3.

Solution: $4^3 = 4 \cdot 4 \cdot 4 = 64$

Example 3: Simplify $(7 - 4)^3 + 3^2 \cdot 5$

Solution:
$$
\begin{aligned}
(7 - 4)^3 + 3^2 \cdot 5 &= 3^3 + 3^2 \cdot 5 &&\text{Simplify inside parentheses.} \\
&= 27 + 9 \cdot 5 &&\text{Write } 3^3 \text{ as 27 and } 3^2 \text{ as 9.} \\
&= 27 + 45 &&\text{Multiply.} \\
&= 72 &&\text{Add.}
\end{aligned}
$$

Example 4: Find the area of the square whose side measures 16 meters.

Solution: Area of a square = (side)^2

= (16 meters)^2

= 256 square meters

The area of the square is 256 square meters.

1.7 EXERCISES

Write using exponential notation.

1. $5 \cdot 5 \cdot 5 \cdot 5 \cdot 5 \cdot 5$

2. $2 \cdot 2 \cdot 2 \cdot 2 \cdot 2 \cdot 2 \cdot 2 \cdot 2$

3. $8 \cdot 8 \cdot 3 \cdot 3 \cdot 3$

4. $7 \cdot 4 \cdot 4 \cdot 9 \cdot 9 \cdot 9 \cdot 9 \cdot 9$

Evaluate.

5. 7^2

6. 8^3

7. 5^4

8. 2^9

9. 3^6

10. 7^4

11. 10^6

12. 9^4

13. 2180^1

14. $17 + 4 \cdot 8$

15. $9 \cdot 5 - 12$

16. $0 \div 10 + 8 \cdot 3$

17. $(11 + 5^2) \div 6$

18. $(10 - 3) \cdot (23 - 19)$

19. $(24 \div 8) + [(4 + 2) \cdot 3]$

20. $42 - \{7 + 4[6 \cdot (12 - 9)] - 60\}$

1.8 INTRODUCTION TO VARIABLES AND ALGEBRAIC EXPRESSIONS

Example 1: Evaluate $x + 5$ if x is 9.

Solution: $x + 5 = 9 + 5$ Replace x with 9.

= 14   Add.

Example 2: Evaluate $7(x - y)$ for $x = 12$ and $y = 8$.

Solution: $7(x - y) = 7(12 - 8)$ Replace x with 12 and y with 8.

= 7(4)   Subtract.

= 28   Multiply.

Example 3: Evaluate $x^3 + z - 2$ for $x = 4$ and $z = 3$.

Solution: $\begin{aligned} x^3 + z - 2 &= 4^3 + 3 - 2 &&\text{Replace } x \text{ with 4 and } z \text{ with 3.} \\ &= 64 + 3 - 2 &&\text{Evaluate } 4^3. \\ &= 65 &&\text{Add and subtract from left to right.} \end{aligned}$

Example 4: Write as a variable expression. Use x to represent "a number".

21 decreased by a number

Solution:

Words	21	decreased by	a number
Translate	21	–	x

$$21 - x$$

1.8 EXERCISES

Evaluate the following expressions when $x = 3$, $y = 6$, and $z = 4$.

1. $4 + 3y$

2. $2xy - 5z$

3. $z + 3x - y$

4. $2xy^2 + 8$

5. $12 - (y - x)$

6. $19 + (3x - 5)$

7. $x^3 + z^2 - y$

8. $z^2 - (x + y)$

9. $\dfrac{2xy}{z}$

10. $\dfrac{8xyz}{9}$

11. $\dfrac{3y + 2}{10}$

12. $\dfrac{7z}{2} + \dfrac{18}{y}$

13. $4z(x + y)$

14. $(xz - 5)^2$

Write the following phrases as variable expressions. Use x to represent "a number".

15. The sum of a number and twelve.

16. A number less twenty-four.

17. A number divided by eight.

18. The difference of a number and one hundred.

19. Eight increased by the product of four and a number.

20. The difference of eleven and a number is added to the quotient of a number and four.

CHAPTER 1 PRACTICE TEST

Evaluate.

1. $48 + 71$

2. $501 - 398$

3. 281×26

4. $41,605 \div 72$

5. $3^4 \cdot 4^2$

6. $7^1 \cdot 3^2$

7. $101 \div 1$

8. $0 \div 52$

9. $29 \div 0$

10. $(7^2 - 6) \cdot 4$

11. $18 + 10 \div 2 \cdot 3 - 9$

12. $3[(7 - 5)^2 + (31 - 27)^2] \div 6$

13. Round 428,791 to the nearest ten-thousand.

Estimate the sum or difference by rounding each number to the nearest hundred.

14. $5176 + 3428 + 769$

15. $7351 - 2067$

Solve.

16. Thirty-one cans of paint cost \$713. How much was each can?

17. Admission to a play costs \$18 per ticket. The Science Club has 12 members who are going to a play together. What is the total cost of their tickets?

18. Henry is looking at two new stoves for his house. One costs \$489 and the other costs \$641. How much more expensive is the higher-priced one?

19. Evaluate $2(x^4 - 7)$ if $x = 3$.

20. Evaluate $\dfrac{4x - 18}{3y}$ if $x = 9$ and $y = 2$.

21. Translate the following phrases into mathematical expressions. Use x to represent "a number".

     (a) The difference of a number and 28.
     (b) Three more than twice a number.

Find the perimeter and the area of each figure.

22.

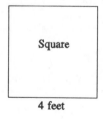

Square

4 feet

23.

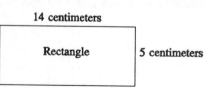

14 centimeters

Rectangle   5 centimeters

The following table shows SAT average scores by state for some selected states. Use the table to answer the following questions.

SAT Average Scores by State

Source: The College Board

	1997		1998		1999	
	Verbal	Math	Verbal	Math	Verbal	Math
Michigan	557	566	558	569	557	565
Minnesota . . .	582	592	585	598	586	598
Mississippi . . .	567	551	562	549	563	548
Missouri	567	568	570	573	572	572
Montana	545	548	543	546	545	546
Nebraska	562	564	565	571	568	571

24. Find the average 1998 SAT math score for students from the state of Montana.

25. Find the <u>increase</u> in average SAT verbal scores for students from the state of Minnesota for the years 1997 and 1999.

CHAPTER 1 HELPFUL HINTS AND INSIGHTS

● The name of the ones period is not used when reading and writing whole numbers.

● A comma may or may not be inserted in a four digit number.

● One way to remember the meaning of the symbols < and > is to think of them as arrowheads "pointing" toward the smaller number.

● A nonzero number divided by 0 is undefined.

● An exponent applies only to its base.

2.1 INTRODUCTION TO INTEGERS

Example 1: Insert < or > between the pair of numbers to make a true statement.

$$-15 \quad -13$$

Solution: −15 is to the left of −13 on a number line, so −15 < −13.

Example 2: Simplify $|-8|$.

Solution: $|-8| = 8$, because −8 is 8 units from 0.

Example 3: Find the opposite of −20.

Solution: The opposite of −20 is $-(-20) = 20$.

Example 4: Simplify $-|-12|$.

Solution: $-|-12| = -12$. The absolute value of −12 is 12 and the opposite of 12 is −12.

2.1 EXERCISES

Represent each quantity by an integer.

1. The Wilson Hornets football team lost 18 yards on a play.

2. The temperature is 75 degrees Fahrenheit above zero.

Graph each integer in the list on the same number line.

3. −4, −1, 2, 3   4. −5, −3, 0, 4

Insert < or > between each pair of integers to make a true statement.

5. −8   3   6. −21   −22   7. 0   −4   8. −19   19

Simplify.

9. $|-98|$ 10. $|22|$ 11. $|-16|$

Find the opposite of each integer.

12. 17   13. −30   14. 101

Simplify.

15. $-|28|$ 16. $-|-17|$ 17. $-(-44)$

Insert < , >, or = between each pair of numbers to make a true statement.

18. $-|-11|$ $-|-10|$ 19. $-|26|$ $-|-26|$ 20. −51 $-(-54)$

2.2 ADDING INTEGERS

Example 1: Add $-5 + (-16)$.

Solution: *Step* 1 $|-5| = 5$, $|-16| = 16$, and $5 + 16 = 21$.
        Step 2   Their common sign is negative, so the sum is negative:
           $-5 + (-16) = -21$

Example 2: Add $7 + (-4)$.

Solution: *Step* 1 $|7| = 7$, $|-4| = 4$, and $7 - 4 = 3$.
        Step 2   7 has the larger absolute value and its sign is $+$:
           $7 + (-4) = +3$ or 3

Example 3: Add $2 + (-12) + (-6) + 10$.

Solution:
$$2 + (-12) + (-6) + 10 = -10 + (-6) + 10$$
$$= -16 + 10$$
$$= -6$$

Example 4: Evaluate $x + y$ if $x = -9$ and $y = 12$.

Solution:
$$x + y = -9 + 12$$
$$= 3$$

2.2 EXERCISES

Add.

1. $21 + 14$

2. $-8 + (-1)$

3. $61 + (-61)$

4. $13 + (-6)$

5. $17 + (-22)$

6. $-14 + (-31)$

7. $-14 + 5 + (-8)$

8. $16 + (-4) + (-5)$

9. $19 + 7 + (-6) + (-22)$

10. $(-13) + 12 + (-11) + 10$

11. $-99 + (-81)$

12. $117 + (-54)$

13. $-87 + 19$

14. $-6 + (-10) + 18 + (-2)$

Evaluate $x + y$ given the following replacement values.

15. $x = -15$ and $y = 14$

16. $x = -60$ and $y = -55$

17. $x = 21$ and $y = -33$

18. $x = -2$ and $y = -82$

Solve.

19. The temperature at 3 p.m. was $-12°$ Celsius. By 10:30 p.m. the temperature had fallen 8 degrees. Find the temperature at 10:30 p.m.

20. Suppose a deep-sea diver dives from the surface to 175 feet below the surface. If the diver swims down 18 feet more, find his depth.

2.3 SUBTRACTING INTEGERS

Example 1: Subtract: $-12 - 5$

Solution: $-12 - 5 = -12 + (-5) = -17$

Example 2: Simplify: $14 - 9 - (-4) - 2$

Solution:
$$14 - 9 - (-4) - 2 = 14 + (-9) + 4 + (-2)$$
$$= 5 + 4 + (-2)$$
$$= 9 + (-2)$$
$$= 7$$

Example 3: Evaluate $x - y$ if $x = -13$ and $y = 21$.

Solution:
$$x - y = -13 - 21$$
$$= -13 + (-21)$$
$$= -34$$

2.3 EXERCISES

Perform indicated operations.

1. $-17 - (-17)$

2. $8 - 5$

3. $11 - 16$

4. $-7 - (-9)$

5. $9 - 10$

6. $-12 - 52$

7. Subtract 20 from -33.

8. Find the difference of -40 and -45.

Simplify.

9. $42 - 35 - 7$

10. $-8 - 9 - (-12)$

11. $14 - (-16) + 8$

12. $-(-13) - 17 + 8$

13. $-20 - 20 - 4$

14. $-20 - (-20) - 4$

Evaluate $x - y$ given the following replacement values.

15. $x = -17$ and $y = 11$

16. $x = -39$ and $y = -68$

Evaluate each expression for the given replacement values.

17. $x - y + z$ if $x = -7$, $y = 6$, and $z = 12$

18. $-x - y + z$ if $x = 40$, $y = -50$ and $z = -60$

Solve.

19. Bobby Williams has $250 in his checking account. He writes a check for $98, makes a deposit of $39, and then writes another check for $111. Find the amount left in his account.

20. In canasta, it is possible to have a negative score. If Omar's score is 20, what is his new score if he loses 30 points.?

2.4 MULTIPLYING AND DIVIDING INTEGERS

Example 1: Multiply: $-4(-8)$

Solution: $-4(-8) = 32$

Example 2: Multiply: $5(-3)(-1)$

Solution: $5(-3)(-1) = (-15)(-1) = 15$

Example 3: Evaluate xy if $x = -10$ and $y = 4$.

Solution: $xy = -10 \cdot 4 = -40$

Example 4: Divide: $\dfrac{-42}{7}$

Solution: $\dfrac{-42}{7} = -6$

Example 5: Evaluate: $(-9)^2$

Solution: $(-9)^2 = (-9)(-9) = 81$

2.4 EXERCISES

Multiply or divide the following.

1. $(-8)(-11)$

2. $0(-23)$

3. $(-21)(2)$

4. $\dfrac{-20}{-5}$

5. $\dfrac{54}{-6}$

6. $\dfrac{101}{0}$

7. $(4)(-3)(-2)$

8. $(-6)(-10)(-3)$

9. $(-1)(5)(-4)(-1)$

10. $(-4)^3$

11. $(-8)^2$

12. $(-39)(13)$

Evaluate ab for the given replacement values.

13. $a = 4$ and $b = -4$

14. $a = -20$ and $b = -6$

15. $a = 80$ and $b = -3$

Evaluate $\dfrac{x}{y}$ for the given replacement values.

16. $x = 78$ and $y = -39$

17. $x = -121$ and $y = -11$

18. $x = 0$ and $y = 99$

Solve.

19. A football team lost 6 yards on each of 3 consecutive plays. Represent the total loss as a product of integers, and find the total loss.

20. Wilhelm lost $200 on each of 6 consecutive days in the stock market. Represent his total loss as a product of integers and find his total loss.

2.5 ORDER OF OPERATIONS

Example 1:   Find the value of each expression.
         (a) $(-6)^2$ (b) -6^2

Solution: (a) $(-6)^2 = (-6)(-6) = 36$
           (b) $-6^2 = -(6 \cdot 6) = -36$

Example 2:   Simplify: $14 + 20 + (-3)^2$

Solution: $14 + 20 + (-3)^2 = 14 + 20 + 9$ Simplify expressions with exponents.
                           $= 43$ Add left to right.

Example 3:   Simplify: $(-2) \cdot |-7| - (-3) + 5^2$

Solution: $(-2) \cdot |-7| - (-3) + 5^2$
         $= (-2) \cdot 7 - (-3) + 5^2$ Write $|-7|$ as 7.
         $= (-2) \cdot 7 - (-3) + 25$ Write 5^2 as 25.
         $= -14 - (-3) + 25$ Multiply.
         $= -11 + 25$ Add or subtract from left to right.
         $= 14$

Example 4:   Evaluate $x - y^2$ if $x = -4$ and $y = 8$.

Solution: $x - y^2 = -4 - 8^2$ Replace x with -4 and y with 8.
               $= -4 - 64$ Simplify.
               $= -68$

2.5 EXERCISES

Simplify.

1. $4 + (-15) \div 3$

2. $9 + 5(4)$

3. $11(-3) + 8$

4. $15 + 4^3$

5. $\dfrac{17 - 8}{-9}$

6. $\dfrac{77}{-5 + 12}$

7. $[5 + (-3)]^4$

8. $50 - (-6)^2$

9. $|7 + 4| \cdot (-8)^2$

10. $6 \cdot 3^2 + 15$

11. $9 + 3^2 - 4^3$

12. $(8 - 16) \div 4$

13. $(24 - 28) \div (31 - 35)$

14. $(-30 - 6) \div 12 - 11$

15. $-4^2 - (-9)^2$

16. $2(-15) \div [4(-7) - 9(-3)]$

17. $\dfrac{(-5)(-6) - (7)(6)}{2[9 \div (10 - 13)]}$

18. $\dfrac{40(-1) - (-4)(-5)}{3[-10 + (-7 + 2)]}$

Evaluate the following expressions if $x = -3$, $y = 5$ and $z = -1$.

19. $3x - y^2$

20. $\dfrac{10x}{z} - 4y$

CHAPTER 2 PRACTICE TEST

Simplify each expression.

1. $17 - 36$

2. $-15 + 9$

3. $4 \cdot (-25)$

4. $(-18) \div (-6)$

5. $(-28) + (-14)$

6. $-11 - (-21)$

7. $(-20) \cdot (-4)$

8. $\dfrac{-108}{-12}$

9. $|-52| + (-17)$

10. $35 - |-100|$

11. $|7| \cdot |-8|$

12. $\dfrac{|-15|}{-|-3|}$

13. $(-12) + 8 + (-4)$

14. $-6 + (-51) - 13 + 7$

15. $(-4)^3 - 42 \div (-7)$

16. $(3 - 10)^2 \cdot (9 - 11)^3$

17. $-(-6)^2 \div 12 \cdot (-5)$

18. $24 - (18 - 16)^3$

19. $-9 + (-39) \div (-13)$

20. $\dfrac{6}{3} - \dfrac{9^2}{27}$

21. $\dfrac{-7(-3) + 13}{-1(-11 - 6)}$

22. $\dfrac{|45 - 50|^2}{8(-3) + 19}$

Evaluate the following when $x = 0$, $y = -4$ and $z = 3$.

23. $5x - y$

24. $|x| - |y| - |z|$

25. $\dfrac{12z}{-3y}$

26. $9 - y$

Solve.

27. A mountain climber is at an elevation of 15,947 feet and moves down the mountain a distance of 6161 feet. Represent his final elevation as a sum and find the sum.

28. Peter Stork has $318 in his checking account. He writes a check for $219, he writes another check for $39, and he deposits $72. Represent the balance in his account by a signed number.

CHAPTER 2 HELPFUL HINTS AND INSIGHTS

- The symbol " – " has two meanings.
    1. It can be read as "minus" and means the operation of subtraction.
    2. It can be read as "negative" and means that the number is less than 0.

- When simplifying expressions with exponents notice that parentheses make an important difference. For example, $(-3)^2$ and -3^2 do not mean the same thing.

- Proper use of order of operations is necessary when simplifying expressions.

3.1 SIMPLIFYING ALGEBRAIC EXPRESSIONS

Example 1: Find the numerical coefficient of the variable term, $-15xy^2$.

Solution: The numerical coefficient of $-15xy^2$ is -15.

Example 2: Simplify: $8y - 9 + 3y + 6$

Solution:
$$\begin{aligned}
8y - 9 + 3y + 6 &= 8y + (-9) + 3y + 6 \\
&= 8y + 3y + (-9) + 6 && \text{Commutative property of addition.} \\
&= (8 + 3)y + (-9) + 6 && \text{Distributive property.} \\
&= 11y - 3 && \text{Simplify.}
\end{aligned}$$

Example 3: Use the distributive property to multiply: $7(x + 5)$

Solution:
$$\begin{aligned}
7(x + 5) &= 7 \cdot x + 7 \cdot 5 && \text{Distributive property.} \\
&= 7x + 35 && \text{Multiply.}
\end{aligned}$$

Example 4: Simplify: $9(x - 2) + 3(4x + 1)$

Solution:
$$\begin{aligned}
&9(x - 2) + 3(4x + 1) \\
&= 9(x) + 9(-2) + 3(4x) + 3(1) && \text{Distributive property.} \\
&= 9x - 18 + 12x + 3 && \text{Multiply.} \\
&= 9x + 12x - 18 + 3 && \text{Rearrange terms.} \\
&= 21x - 15 && \text{Combine like terms.}
\end{aligned}$$

Example 5: Find the area of the rectangular garden.

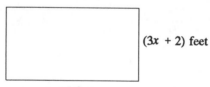

$(3x + 2)$ feet

15 feet

Solution:
$$\begin{aligned}
A &= \text{length} \cdot \text{width} \\
&= 15(3x + 2) && \text{Let length} = 15 \text{ and width} = 3x + 2. \\
&= 45x + 30 && \text{Multiply.}
\end{aligned}$$

The area is $(45x + 30)$ square feet.

3.1 EXERCISES

Find the numerical coefficient of each variable term.

1. $22d$ 2. $-y$ 3. $14x^3y^2$

Simplify the following by combining like terms.

4. $13x - 4x$ 5. $3b - 7b - b$ 6. $12y + 9y - 2y + 7$

Muliply.

7. $-5(11x)$ 8. $2(3x - 1)$ 9. $7(4x + 8)$

Simplify the following.

10. $4(x - 11) + 16$

11. $2(9 - 3a) - 16a$

12. $-5(2x + 7) + 4(x - 6)$

13. $-x + 13y - 9x + 18y$

14. $-12(1 + 2v) + 4v$

15. $-4(m - 9) + 8(m + 3)$

16. $22x + 6(2x - 5)$

17. $9(4xy + 3) - 1(2xy - 8)$

Find the perimeter:

18.

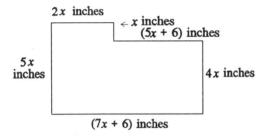

Find the area of each figure.

19.

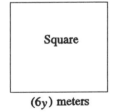

20.

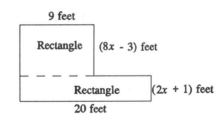

3.2 SOLVING EQUATIONS: THE ADDITION PROPERTY

Example 1: Is -3 a solution of the equation $8(x - 2) = 40$?

Solution:
$$8(x - 2) = 40$$
$$8(-3 - 2) = 40 \quad ? \qquad \text{Replace } x \text{ with } -3.$$
$$8(-5) = 40 \quad ?$$
$$-40 = 40 \quad ? \qquad \text{False.}$$

Since $-40 = 40$ is false, -3 is **not** a solution of the equation.

Example 2: Solve for x: $x - 5 = 9$

Solution:
$$x - 5 = 9$$
$$x - 5 + 5 = 9 + 5 \qquad \text{Add 5 to both sides of the equation.}$$
$$x = 14 \qquad \text{Simplify.}$$

To check, replace x with 14 in the original equation.

$$x - 5 = 9 \qquad \text{Original equation.}$$
$$14 - 5 = 9 \qquad \text{Replace } x \text{ with 14.}$$
$$9 = 9 \qquad \text{True.}$$

Since $9 = 9$ is a true statement, 14 is the solution of the equation.

Example 3: Solve for y: $y - 7 = -1 + 11$

Solution:
$$y - 7 = -1 + 11$$
$$y - 7 = 10 \qquad \text{Simplify the right side of the equation.}$$
$$y - 7 + 7 = 10 + 7 \qquad \text{Add 7 to both sides.}$$
$$y = 17 \qquad \text{Simplify.}$$

Check to see that 17 is the solution of the equation.

Example 4: Solve for x: $3x + 9 - 2x = 18 - 22$

Solution:
$$3x + 9 - 2x = 18 - 22$$
$$3x - 2x + 9 = 18 - 22 \qquad \text{Simplify \underline{each} side of the equation separately.}$$
$$1x + 9 = -4$$
$$1x + 9 - 9 = -4 - 9 \qquad \text{Subtract 9 from each side.}$$
$$1x = -13$$
$$\text{or} \quad x = -13$$

Check to verify that -13 is the solution.

3.2 EXERCISES

Decide whether the given number is a solution to the given equation.

1. Is 8 a solution to $y + 9 = 17$?

2. Is -4 a solution to $x - 12 = -16$?

3. Is 0 a solution to $3(a - 2) = 6$?

4. Is -3 a solution to $5b = 47 - b$?

Solve the following.

5. $x + 7 = 23$

6. $t - 14 = 21$

7. $x - 6 = 2 + 8$

8. $-9 - 2 = -5 + y$

9. $4a + 3 - 3a = -2 + 12$

10. $11 - 11 = 9x - 8x$

11. $20 + (-17) = 10x + 7 - 9x$

12. $78 = w + 78$

13. $y + 4 = -14$

14. $m + 20 = 22$

15. $-33 + x = -35$

16. $-17n - 10 + 18n = -6$

17. $5y + 9 - 4y = 42$

18. $-19x + 13 + 20x = -1 - 5$

19. $31x + 27 - 30x = -27$

20. $z - 82 = -25$

3.3 SOLVING EQUATIONS: THE MULTIPLICATION PROPERTY

Example 1: Solve for x: $-6x = 48$

Solution:

$-6x = 48$ Original equation.

$$\frac{-6x}{-6} = \frac{48}{-6}$$ Divide both sides by -6.

$$\frac{-6}{-6} \cdot x = \frac{48}{-6}$$

$1 \cdot x = -8$ Simplify.
or $x = -8$

To check, replace x with -8 in the orginal equation.

$-6x = 48$ Original equation.
$-6(-8) = 48$ Let $x = -8$.
$48 = 48$ True.

The solution is -8.

Example 3: Solve for m: $14m - 19m = -20 + 45$

Solution:

$14m - 19m = -20 + 45$
$-5m = 25$ Combine like terms on each side of the equation.

$$\frac{-5m}{-5} = \frac{25}{-5}$$ Divide both sides by -5.

$m = -5$ Simplify.

Check to see that the solution is -5.

3.3 EXERCISES

Solve the following equations.

1. $7x = 56$

2. $-9z = 99$

3. $-5y = -105$

4. $2m = 24$

5. $4b - 9b = -35$

6. $18 = 6t - 3t$

7. $3x - 8x = -10 + (-10)$

8. $-12y = 12$

9. $n + 6n = 77$

23

10. $-8x = -8$ 11. $y - 9y = 48$ 12. $14a - 6a = 32$

13. $-64 + 26 = -11x - 8x$ 14. $13v + 2v = 45$ 15. $11y - 5y = -6 - 60$

16. $32b - 34b = 82$

Translate each phrase to an algebraic expression.

17. Eight added to the product of 9 and a number.

18. Three times a number decreased by 20.

19. The quotient of 55 and the product of a number and -4.

20. Twice the difference of a number and -6.

3.4 SOLVING LINEAR EQUATIONS IN ONE VARIABLE

Example 1: Solve: $4x - 9 = 15$

Solution:
$$4x - 9 = 15$$
$$4x - 9 + 9 = 15 + 9 \qquad \text{Add 9 to both sides.}$$
$$4x = 24 \qquad \text{Simplify.}$$

$$\frac{4x}{4} = \frac{24}{4} \qquad \text{Divide both sides by 4.}$$
$$x = 6 \qquad \text{Simplify.}$$

Check:

$$4x - 9 = 15$$
$$4(6) - 9 = 15 \qquad \text{Replace } x \text{ with 6 and simplify.}$$
$$24 - 9 = 15$$
$$15 = 15 \qquad \text{True.}$$

The solution is 6.

Example 2: Solve: $5y - 8 = 2y + 13$

Solution:
$$5y - 8 = 2y + 13$$
$$5y - 8 + 8 = 2y + 13 + 8 \qquad \text{Add 8 to both sides.}$$
$$5y = 2y + 21 \qquad \text{Simplify.}$$
$$5y - 2y = 2y + 21 - 2y \qquad \text{Subtract } 2y \text{ from both sides.}$$
$$3y = 21 \qquad \text{Simplify.}$$

$$\frac{3y}{3} = \frac{21}{3} \qquad \text{Divide both sides by 3.}$$

$$y = 7 \qquad \text{Simplify.}$$

Check to see that the solution is 7.

Example 3: Solve: $8(x - 1) = 10x + 12$

Solution:

$$8(x - 1) = 10x + 12$$
$$8x - 8 = 10x + 12 \qquad \text{Distributive property.}$$
$$8x - 8 + 8 = 10x + 12 + 8 \qquad \text{Add 8 to both sides.}$$
$$8x = 10x + 20 \qquad \text{Simplify.}$$
$$8x - 10x = 10x + 20 - 10x \qquad \text{Subtract } 10x \text{ from both sides.}$$
$$-2x = 20 \qquad \text{Simplify.}$$

$$\frac{-2x}{-2} = \frac{20}{-2} \qquad \text{Divide both sides by } -2.$$

$$x = -10 \qquad \text{Simplify.}$$

Check to see that -10 is the solution.

Example 4: Solve: $9(4 - x) + 18 = 0$

Solution:

$$9(4 - x) + 18 = 0$$
$$36 - 9x + 18 = 0 \qquad \text{Distributive property.}$$
$$-9x + 54 = 0 \qquad \text{Combine like terms on the left side of the equation.}$$
$$-9x + 54 - 54 = 0 - 54 \qquad \text{Subtract 54 from both sides.}$$
$$-9x = -54 \qquad \text{Simplify.}$$
$$\frac{-9x}{-9} = \frac{-54}{-9} \qquad \text{Divide both sides by } -9.$$
$$x = 6 \qquad \text{Simplify.}$$

Check to see that 6 is the solution.

3.4 EXERCISES

Solve each equation.

1. $5x - 30 = 0$

2. $2b - 6 = 8$

3. $7y - 4 = 3y + 12$

4. $9 - a = 11$

5. $6d + 25 = 31$

6. $8y - 3 = -19$

7. $4p - 28 = 0$

8. $5t + 2 = 37$

9. $9x + 7 = 6x - 8$

10. $-y - 11 = 5y - 17$

11. $-11x + 1 = -12x + 6$

12. $14 - 5y = 14 + 2y$

13. $3(m - 2) = m - 10$

14. $7(4c + 2) - 1 = 26c + 5$

15. $9(4 - x) = 4(x - 4)$

16. $20 + 8(y - 1) = 6y + 32$

Write each sentence as an equation.

17. The sum of -24 and 13 is -11.

18. Four times the difference of -10 and 3 amounts to -52.

19. Twenty subtracted from -15 equals -35.

20. The quotient of 150 and twice 5 is equal to 15.

3.5 LINEAR EQUATIONS IN ONE VARIABLE AND PROBLEM SOLVING

Example 1: Write the following sentences as equations. Use x to represent "a number".

    (a) Ten more than a number is 16.

    (b) Three times a number is -21.

Solution: (a) In words: Ten   more than   a number   is   16

             Translate: 10   + x =   16

    (b) In words: Three   times   a number   is -21

             Translate: 3   · x = -21

$$3x = -21$$

Example 2: Twice a number decreased by 4 is the same as the number increased by 7. Find the number.

Solution: Let x = the unknown number

In words: Twice a number   decreased by   4   is the same as   the number   increased by   7
                $2x$ $-$ 4   = x $+$ 7

Translate:

$$2x - 4 = x + 7$$
$$2x - 4 + 4 = x + 7 + 4 \qquad \text{Add 4 to both sides.}$$
$$2x = x + 11 \qquad \text{Simplify.}$$
$$2x - x = x + 11 - x \qquad \text{Subtract } x \text{ from both sides}$$
$$x = 11 \qquad \text{Simplify.}$$

Check: Twice "11" is 22 and 22 decreased by 4 is 18. This is the same as "11" increased by 7.

The unknown number is 11.

Example 3: Dan McGill sold a used stereo system and a CD collection for $350, receiving six times as much money for the stereo system as for the CD collection. Find the price of each.

Solution: Let x = CD collection price. Then
           $6x$ = the stereo system price

In words: stereo price   +   CD price   is   350
Translate: $6x$ + x =   350

$$7x = 350 \qquad \text{Combine like terms.}$$

$$\frac{7x}{7} = \frac{350}{7} \qquad \text{Divide both sides by 7.}$$

$$x = 50 \qquad \text{Simplify.}$$

Check: The CD collection sold for $50. The stereo system sold for $6x = 6(\$50) = \$300$. Since $\$50 + \$300 = \$350$, the total price, and $300 is six-times $50, the solution checks.

The CD collection sold for $50 and the stereo system sold for $300.

3.5 EXERCISES

Write the sentences as equations. Use x to represent "a number".

1. A number added to −8 is −12.

2. Four times a number yields 44.

3. Three added to twice a number gives −11.

4. Seven times the difference of 6 and a number amounts to −14.

Solve.

5. Five times a number added to twelve is thirty-seven. Find the number.

6. The product of eight and a number gives fifty six. Find the number.

7. A number less nine is twelve. Find the number.

8. Fourteen decreased by some number equals the quotient of twenty and four. Find the number.

9. The sum of four, five and a number is ten. Find the number.

10. The product of thirteen and a number is one hundred sixty-nine. Find the number.

11. Seventeen added to the product of five and some number amounts to the product of seven and the same number added to twenty-five. Find the number.

12. Seventy-three less a number is equal to the product of six and the sum of the number and four. Find the number.

13. Billy and Bob Mazurk collect football cards. Billy has three times the number of cards Bob has. Together they have 960 cards. Find how many cards Billy has.

14. A boat and trailer are worth $7800. The boat is worth five times as much money as the trailer. Find the value of the boat and the value of the trailer.

15. Tonya O'Brien sold her doll collection and accessories for $560. If she received six times as much money for the dolls as she did for the accessories, find how much money she received for the dolls.

16. Doug's car is traveling twice as fast as Karen's car. If their combined speed is 90 miles per hour, find the speed of Doug's car.

CHAPTER 3 PRACTICE TEST

1. Simplify $4x - 7 - 9x + 12$ by combining like terms.

2. Multiply: $-5(4y - 5)$

3. Simplify: $7(2z + 9) - 3z + 21$

4. Find the perimeter.

Square | $(5x + 8)$ meters

5. Find the area.

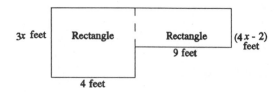

Rectangle — $3x$ feet, 4 feet
Rectangle — 9 feet, $(4x - 2)$ feet

Solve the following equations.

6. $7x + x = -64$

7. $29 = 2x - 31x$

8. $5b - 7 = 28$

9. $14 + 6z = 50$

10. $11x + 21 - 10x - 18 = 30$

11. $1 - d + 6d = 31$

12. $6x - 9 = -57$

13. $-9y + 8 = -10$

14. $4(x - 3) = 0$

15. $6(3 + 8y) = 18$

16. $11x - 5 = x + 25$

17. $14a - 3 = 5a + 24$

18. $4 + 5(2m - 3) = 9$

19. $8(2x + 5) = 10(x + 4)$

Write each sentence as an equation.

20. The product of -18 and 3 yields -54.

21. Twice the difference of 12 and -5 amounts to 34.

Solve the following.

22. The sum of four times a number and six times the same number is thirty. Find the number.

23. Thirty-two less half of 32 is equal to the sum of some number and 5. Find the number.

24. In a local basketball game, Gary made three times as many free throws as Jamal. If the total number of free throws made by both men was twenty-four, find how many free throws Gary made.

25. In a bowling league, there are eighteen more men than women. Find the number of women in the league if the total number of league members is 90.

CHAPTER 3 HELPFUL HINTS AND INSIGHTS

- It does not matter on which side of the equation the variable is isolated.

- It is always a good idea to check the solution of an equation that we have solved to see that it makes the equation a true statement.

- When solving an equation check to see whether a side of the equation can be simplified before applying a property of equality.

4.1 INTRODUCTION TO FRACTIONS AND EQUIVALENT FRACTIONS

Example 1: Graph $\frac{5}{4}$ on a number line.

Solution:

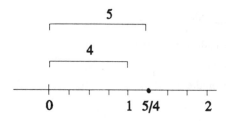

Example 2: Write $\frac{3}{8}$ as an equivalent fraction whose denominator is 32.

Solution: Since $8 \cdot 4 = 32$, multiply the numerator and denominator by 4.

$$\frac{3}{8} = \frac{3 \cdot 4}{8 \cdot 4} = \frac{12}{32}$$

Then $\frac{3}{8}$ is equivalent to $\frac{12}{32}$.

Example 3: Write $\frac{6}{7x}$ as an equivalent fraction whose denominator is $14x$.

Solution: $7x \cdot 2 = 14x$. Multiply the numerator and denominator by 2.

$$\frac{6}{7x} = \frac{6 \cdot 2}{7x \cdot 2} = \frac{12}{14x}$$

Example 4; Simplify $\frac{16}{-4}$ by dividing the numerator by the denominator.

Solution: $\frac{16}{-4} = -4$

4.1 EXERCISES

Represent the shaded part of each geometric figure by a proper fraction.

1.

2.

Represent the shaded part of the geometric figures by an improper fraction.

3.

4.

Graph the fraction on a number line.

5. $\dfrac{2}{3}$

6. $\dfrac{5}{7}$

7. $\dfrac{9}{5}$

Write each fraction as an equivalent fraction with the given denominator.

8. $\dfrac{7}{9}$; denominator of 45

9. $\dfrac{4}{5}$; denominator of 60

10. $\dfrac{6a}{11}$; denominator of 33

11. $\dfrac{8y}{13}$; denominator of 26

12. $\dfrac{4}{9b}$; denominator of $36b$

13. 6: denominator of 9

14. 2; denominator of 15

Simplify by dividing.

15. $\dfrac{11}{11}$

16. $\dfrac{-9}{-9}$

17. $\dfrac{27}{9}$

18. $\dfrac{-36}{12}$

19. $\dfrac{12}{-12}$

20. $\dfrac{30}{1}$

4.2 FACTORS AND SIMPLEST FORM

Example 1: Write the prime factorization of 48.

Solution: $48 = 2 \cdot 24$

$= 2 \cdot 2 \cdot 12$

$= 2 \cdot 2 \cdot 2 \cdot 6$

$= 2 \cdot 2 \cdot 2 \cdot 2 \cdot 3$

$48 = 2 \cdot 2 \cdot 2 \cdot 2 \cdot 3$ or $2^4 \cdot 3$

Example 2: Write the prime factorization of 2052.

$$2052 = 2 \cdot 1026$$
$$= 2 \cdot 2 \cdot 513$$

Solution:
$$= 2 \cdot 2 \cdot 3 \cdot 171$$
$$= 2 \cdot 2 \cdot 3 \cdot 3 \cdot 57$$
$$= 2 \cdot 2 \cdot 3 \cdot 3 \cdot 3 \cdot 19$$

$$2052 = 2 \cdot 2 \cdot 3 \cdot 3 \cdot 3 \cdot 19 \quad \text{or} \quad 2^2 \cdot 3^3 \cdot 19$$

Example 3: Simplify: $\dfrac{14}{56}$

Solution: First, write the prime factorization of the numerator and the denominator.

$$\frac{14}{56} = \frac{2 \cdot 7}{2 \cdot 2 \cdot 2 \cdot 7} = \frac{1}{4}$$

Example 4: Simplify: $\dfrac{8y}{20y}$

Solution: $\dfrac{8y}{20y} = \dfrac{2 \cdot 2 \cdot 2 \cdot y}{2 \cdot 2 \cdot 5 \cdot y} = \dfrac{2}{5}$

Example 5: Simplify: $\dfrac{40x^2}{35x}$

Solution: $\dfrac{40x^2}{35x} = \dfrac{2 \cdot 2 \cdot 2 \cdot 5 \cdot x \cdot x}{5 \cdot 7 \cdot x} = \dfrac{8x}{7}$

4.2 EXERCISES

Write the prime factorization of each number.

1. 84   2. 180   3. 120   4. 504

Simplify each fraction.

5. $\dfrac{14}{21}$ 6. $\dfrac{80}{90}$ 7. $\dfrac{15}{35}$ 8. $\dfrac{28}{6}$

9. $\dfrac{70}{42}$ 10. $\dfrac{121}{33}$ 11. $\dfrac{18b}{45b}$ 12. $\dfrac{12xy}{52x}$

13. $\dfrac{19x^2}{38x}$ 14. $\dfrac{51x}{85xy}$ 15. $\dfrac{90m^2}{350mn}$ 16. $\dfrac{8xyz}{54xz}$

17. $\dfrac{144ab}{84b^2}$ 18. $\dfrac{504x}{45x^2}$

19. There are 1760 yards in a mile. What fraction of a mile is represented by 220 yards?

20. John works 40 hours a week. What fraction of his work week is represented by 6 hours?

4.3 MULTIPLYING AND DIVIDING FRACTIONS

Example 1: Multiply: $\dfrac{3}{4} \cdot \dfrac{7}{8}$

Solution: $\dfrac{3}{4} \cdot \dfrac{7}{8} = \dfrac{3 \cdot 7}{4 \cdot 8}$ Product of numerators.
Product of denominators.

$$= \dfrac{21}{32}$$

Example 2: Multiply: $\dfrac{28}{15} \cdot \dfrac{10}{21}$

Solution: $\dfrac{28}{15} \cdot \dfrac{10}{21} = \dfrac{28 \cdot 10}{15 \cdot 21}$

Next, simplify by prime factoring and dividing out common factors.

$$\dfrac{28 \cdot 10}{15 \cdot 21} = \dfrac{2 \cdot 2 \cdot 7 \cdot 2 \cdot 5}{3 \cdot 5 \cdot 3 \cdot 7}$$

$$= \dfrac{8}{9}$$

Example 3: Multiply: $\dfrac{5x}{6} \cdot \dfrac{12}{11x}$

Solution: $\dfrac{5x}{6} \cdot \dfrac{12}{11x} = \dfrac{5 \cdot x \cdot 12}{6 \cdot 11 \cdot x}$

$$= \dfrac{5 \cdot 2 \cdot 2 \cdot 3}{2 \cdot 3 \cdot 11}$$

$$= \dfrac{10}{11}$$

Example 4: Evaluate: $\left(\dfrac{2}{3}\right)^3$

Solution: $\left(\dfrac{2}{3}\right)^3 = \dfrac{2}{3} \cdot \dfrac{2}{3} \cdot \dfrac{2}{3} = \dfrac{2 \cdot 2 \cdot 2}{3 \cdot 3 \cdot 3} = \dfrac{8}{27}$

Example 5: Divide: $\dfrac{3x}{4} \div 8$

Solution: $\dfrac{3x}{4} \div 8 = \dfrac{3x}{4} \div \dfrac{8}{1} = \dfrac{3x}{4} \cdot \dfrac{1}{8} = \dfrac{3x \cdot 1}{4 \cdot 8} = \dfrac{3x}{32}$

Example 6: If $x = \dfrac{6}{7}$ and $y = -\dfrac{1}{5}$, evaluate xy.

Solution:
$$xy = \dfrac{6}{7} \cdot -\dfrac{1}{5}$$
$$= -\dfrac{6 \cdot 1}{7 \cdot 5}$$
$$= -\dfrac{6}{35}$$

4.3 EXERCISES

Perform the following operations.

1. $\dfrac{3}{4} \cdot \dfrac{10}{33}$
2. $-\dfrac{5}{6} \cdot \dfrac{4}{7}$
3. $\dfrac{8y}{11} \cdot \dfrac{22}{18}$
4. $9 \cdot \dfrac{1}{2}$
5. $-\dfrac{5}{6} \cdot 12$

6. $\dfrac{5}{3} \div \dfrac{1}{6}$
7. $\dfrac{20}{33} \div -\dfrac{5}{11}$
8. $\dfrac{12y}{13} \div \dfrac{3}{26}$
9. $-\dfrac{7}{8} \div 14$
10. $\left(-\dfrac{7}{8}\right)^2$

11. $\left(\dfrac{2}{3}\right)^5$
12. $\dfrac{1}{4} \cdot \dfrac{2}{5} \div \dfrac{3}{10}$
13. $\dfrac{4}{9} \div \dfrac{2}{27} \cdot -\dfrac{2}{5}$
14. $12x \div \dfrac{24x}{5}$
15. $\left(25 \div \dfrac{5}{4}\right) \cdot \dfrac{3}{16}$

Given the replacement values, evaluate (a) xy and (b) $x \div y$.

16. $x = \dfrac{3}{7}$ and $y = \dfrac{7}{9}$
17. $x = -\dfrac{8}{15}$ and $y = \dfrac{3}{4}$

Determine whether the replacement values are solutions of the given equations.

18. $2x = -\dfrac{5}{8};\ \ x = -\dfrac{5}{16}$

19. $-\dfrac{3}{4}y = \dfrac{1}{12};\ \ y = \dfrac{1}{9}$

20. If $\dfrac{5}{8}$ of a 64-tree orchard are apple trees, find how many apple trees are in the orchard.

4.4 ADDING AND SUBTRACTING LIKE FRACTIONS
AND LEAST COMMON DENOMINATOR

Example 1: Add: $\dfrac{2}{15} + \dfrac{11}{15}$

Solution: $\dfrac{2}{15} + \dfrac{11}{15} = \dfrac{2 + 11}{15}$ Add numerators.
Same denominator.

$$= \dfrac{13}{15}$$

Example 2: Subtract: $\dfrac{7x}{8} - \dfrac{5}{8}$

Solution: $\dfrac{7x}{8} - \dfrac{5}{8} = \dfrac{7x - 5}{8}$

The terms in the numerator are unlike terms and cannot be combined.

Example 3: Evaluate $x - y$ if $x = -\dfrac{7}{9}$ and $y = \dfrac{1}{9}$.

Solution: $x - y = -\dfrac{7}{9} - \dfrac{1}{9}$ Replace x with $-\dfrac{7}{9}$ and y with $\dfrac{1}{9}$.

$$= \dfrac{-7 - 1}{9}$$

$$= \dfrac{-8}{9} \quad \text{or} \quad -\dfrac{8}{9}$$

Example 4: Find the LCD of $-\dfrac{3}{8}$, $\dfrac{1}{3}$, and $\dfrac{7}{9}$.

Solution:
$8 = 2 \cdot 2 \cdot 2$
$3 = 3$
$9 = 3 \cdot 3$

$\text{LCD} = 2 \cdot 2 \cdot 2 \cdot 3 \cdot 3 = 72$

Example 5: Solve: $x - \dfrac{1}{6} = \dfrac{5}{6}$

Solution: $x - \dfrac{1}{6} = \dfrac{5}{6}$

$x - \dfrac{1}{6} + \dfrac{1}{6} = \dfrac{5}{6} + \dfrac{1}{6}$ Add $\dfrac{1}{6}$ to both sides.

$$x = \dfrac{6}{6}$$

$x = 1$ Reduce.

4.4 EXERCISES

Add or subtract the following.

1. $\dfrac{1}{8} + \dfrac{5}{8}$

2. $\dfrac{5}{12} + \dfrac{7}{12}$

3. $-\dfrac{1}{3} + \dfrac{1}{3}$

4. $-\dfrac{2}{x} + \dfrac{6}{x}$

5. $\dfrac{7}{10y} + \dfrac{1}{10y}$

6. $\dfrac{3}{17} + \dfrac{2}{17} + \dfrac{5}{17}$

7. $\dfrac{9}{13} - \dfrac{5}{13}$

8. $\dfrac{2}{y} - \dfrac{8}{y}$

9. $\dfrac{5}{22} - \dfrac{3}{22}$

10. $\dfrac{8}{15a} + \dfrac{13}{15a}$

11. $-\dfrac{5}{14} + \dfrac{3}{14}$

12. $\dfrac{11x}{16} - \dfrac{17x}{16}$

Determine whether the given value is a solution of the given equation.

13. $x + \dfrac{5}{8} = \dfrac{1}{2}; \quad x = -\dfrac{1}{8}$

14. $-\dfrac{3}{4} + y = -\dfrac{1}{8}; \quad y = \dfrac{1}{4}$

Find the LCD of the list of fractions.

15. $\dfrac{1}{6}, \dfrac{5}{27}$

16. $-\dfrac{3}{8}, \dfrac{7}{36}$

Solve.

17. $x + \dfrac{1}{10} = -\dfrac{7}{10}$

18. $4x + \dfrac{1}{13} - 3x = \dfrac{3}{13} - \dfrac{9}{13}$

19. A recipe for Chocolate Divine cake calls for $\dfrac{4}{3}$ cups of flour and later $\dfrac{2}{3}$ cup of flour. Find how much flour is needed to make the recipe.

20. Sally combined $\dfrac{1}{8}$ of a cup of milk with $\dfrac{4}{8}$ of a cup of milk in a pitcher. She then poured $\dfrac{2}{8}$ of a cup of milk from the pitcher onto her cereal. How much milk was left in the pitcher?

4.5 ADDING AND SUBTRACTING UNLIKE FRACTIONS

Example 1: Add: $\dfrac{3}{16} + \dfrac{5}{8}$

Solution: *Step 1* The LCD for denominators 16 and 8 is 16.

Step 2 $\dfrac{3}{16} = \dfrac{3}{16}, \quad \dfrac{5}{8} = \dfrac{5 \cdot 2}{8 \cdot 2} = \dfrac{10}{16}$

Step 3 Add.

$$\dfrac{3}{16} + \dfrac{5}{8} = \dfrac{3}{16} + \dfrac{10}{16} = \dfrac{13}{16}$$

Example 2: Subtract: $\dfrac{3}{4} - \dfrac{7}{9}$

Solution: *Step* 1   The LCD for denominators 4 and 9 is 36.

     Step 2 $\dfrac{3}{4} = \dfrac{3 \cdot 9}{4 \cdot 9} = \dfrac{27}{36}$ and $\dfrac{7}{9} = \dfrac{7 \cdot 4}{9 \cdot 4} = \dfrac{28}{36}$

     Step 3   Subtract.

$$\frac{3}{4} - \frac{7}{9} = \frac{27}{36} - \frac{28}{36}$$

$$= \frac{27 - 28}{36}$$

$$= \frac{-1}{36} \quad \text{or} \quad -\frac{1}{36}$$

Example 3: Find: $5 - \dfrac{x}{6}$

Solution: Recall that $5 = \dfrac{5}{1}$. The LCD for denominators 1 and 6 is 6.

$$\frac{5}{1} - \frac{x}{6} = \frac{5 \cdot 6}{1 \cdot 6} - \frac{x}{6}$$

$$= \frac{30}{6} - \frac{x}{6}$$

$$= \frac{30 - x}{6}$$

Example 4: Solve: $x + \dfrac{2}{5} = -\dfrac{3}{15}$

Solution: $x + \dfrac{2}{5} = -\dfrac{3}{15}$

$$x + \frac{2}{5} + \left(-\frac{2}{5}\right) = -\frac{3}{15} + \left(-\frac{2}{5}\right)$$

$$x = \frac{3}{15} + \frac{-2 \cdot 3}{5 \cdot 3} \qquad \text{The LCD for denominators 5 and 15 is 15.}$$

$$x = \frac{-3 + (-6)}{15}$$

$$x = \frac{-9}{15} \quad \text{or} \quad -\frac{9}{15}$$

$$x = -\frac{3 \cdot 3}{3 \cdot 5} = -\frac{3}{5} \qquad \text{Write } -\frac{9}{15} \text{ in lowest terms.}$$

4.5 EXERCISES

Perform the indicated operations.

1. $\dfrac{3}{8} + \dfrac{1}{16}$

2. $\dfrac{4}{5} - \dfrac{7}{10}$

3. $-\dfrac{2}{9} + \dfrac{5}{3}$

4. $\dfrac{7x}{13} - \dfrac{5}{26}$

5. $\dfrac{5y}{6} - \dfrac{5}{12}$

6. $9x - \dfrac{3}{10}$

7. $\dfrac{9}{20} + \dfrac{5}{20} + \dfrac{3}{20}$

8. $\dfrac{8}{23} + \dfrac{31}{23} - \dfrac{16}{23}$

9. $-\dfrac{4}{5} + \dfrac{9}{10} - \dfrac{1}{15}$

10. $\dfrac{y}{2} + \dfrac{y}{6} + \dfrac{5y}{12}$

11. $\dfrac{5}{7} + \dfrac{6}{11x}$

12. $\dfrac{4}{15x} + \dfrac{3}{4}$

13. $\dfrac{5}{12} + \dfrac{9b}{16} - \dfrac{7}{8}$

14. $\dfrac{4x}{13} - \dfrac{3x}{26} - \dfrac{5}{2}$

Evaluate the expression when $x = \dfrac{1}{4}$ *and* $y = \dfrac{2}{5}$.

15. $2x + y$

16. $x - y$

Solve each equation.

17. $x - \dfrac{8}{9} = \dfrac{2}{3}$

18. $15y - \dfrac{3}{7} - 14y = \dfrac{13}{14}$

Solve.

19. A freight truck has $\dfrac{2}{5}$ ton of printers, $\dfrac{1}{2}$ ton of copiers, and $\dfrac{1}{3}$ ton of stereos. Find the total weight of its load.

20. Find the difference in length of two boards if one board is $\dfrac{5}{6}$ of a foot long and the other is $\dfrac{3}{4}$ of a foot long.

4.6 COMPLEX FRACTIONS AND REVIEW OF ORDER OF OPERATIONS

Example 1: Simplify: $\dfrac{\dfrac{x}{3}}{\dfrac{5}{6}}$

Solution:
$$\frac{\dfrac{x}{3}}{\dfrac{5}{6}} = \frac{x}{3} \div \frac{5}{6}$$

$$= \frac{x}{3} \cdot \frac{6}{5}$$

$$= \frac{x \cdot 2 \cdot 3}{3 \cdot 5}$$

$$= \frac{2x}{5}$$

Example 2: Simplify:
$$\frac{\dfrac{x}{4} - 5}{\dfrac{3}{8}}$$

Solution:
$$\frac{\dfrac{x}{4} - 5}{\dfrac{3}{8}} = \frac{8\left(\dfrac{x}{4} - 5\right)}{8\left(\dfrac{3}{8}\right)}$$
     Multiply the numerator and denominator by the LCD 8.

$$= \frac{\left(8 \cdot \dfrac{x}{4}\right) - (8 \cdot 5)}{8\left(\dfrac{3}{8}\right)}$$
     Distributive property.

$$= \frac{2x - 40}{3}$$
     Multiply.

Example 3: Simplify:
$$\left(-\frac{1}{6} - \frac{11}{6}\right) \div \frac{7}{12}$$

Solution:
$$\left(-\frac{1}{6} - \frac{11}{6}\right) \div \frac{7}{12} = -\frac{12}{6} \div \frac{7}{12}$$

$$= -\frac{2}{1} \cdot \frac{12}{7}$$

$$= -\frac{24}{7}$$

4.6 EXERCISES

Simplify the following.

1. $\dfrac{\frac{8}{27}}{\frac{1}{9}}$

2. $\dfrac{\frac{2x}{13}}{\frac{4}{3}}$

3. $\dfrac{\frac{3}{8} + \frac{1}{4}}{\frac{2}{5} + \frac{7}{10}}$

4. $\dfrac{\frac{7x}{5}}{6 - \frac{1}{10}}$

5. $\dfrac{\frac{6}{11} + 1}{\frac{3}{8a}}$

6. $\dfrac{9 - \frac{1}{3}}{7 + \frac{3}{5}}$

7. $\left(-\dfrac{3}{8} - \dfrac{9}{8}\right) \div \dfrac{5}{16}$

8. $5^2 - \left(\dfrac{4}{3}\right)^2$

9. $\left(3 - \dfrac{5}{2}\right)^2$

10. $\left(\dfrac{4}{5} - 1\right)\left(\dfrac{3}{4} + \dfrac{7}{8}\right)$

11. $\left(\dfrac{4}{5} \cdot \dfrac{5}{9}\right) - \left(\dfrac{2}{3} \div \dfrac{3}{7}\right)$

12. $\left(-\dfrac{8}{7} + \dfrac{1}{7}\right)^5$

13. $\dfrac{\frac{x}{4} + 5}{2 + \frac{2}{3}}$

14. $\dfrac{7 - \frac{x}{6}}{10 + \frac{5}{12}}$

Evaluate the expression for $x = \dfrac{2}{5}$ and $y = -\dfrac{5}{6}$.

15. $\dfrac{1 + x}{y}$

16. $10x - y$

17. $x^2 + y$

18. xy

Find the average of each pair of numbers.

19. $\dfrac{4}{9}, \ \dfrac{7}{18}$

20. $\dfrac{5}{6}, \ \dfrac{7}{18}$

4.7 SOLVING EQUATIONS CONTAINING FRACTIONS

Example 1: Solve for x: $\dfrac{3}{4}x = 9$

Solution: $\qquad \dfrac{3}{4}x = 9$

$\qquad \dfrac{4}{3} \cdot \dfrac{3}{4}x = \dfrac{4}{3} \cdot 9 \qquad$ Multiply both sides by $\dfrac{4}{3}$.

$\qquad\qquad 1x = 12 \qquad$ Simplify.

$\qquad$ or $\ \ x = 12$

Example 2: Solve for y: $\dfrac{y}{8} + 2 = \dfrac{3}{4}$

Solution:

$$\dfrac{y}{8} + 2 = \dfrac{3}{4}$$

$$8\left(\dfrac{y}{8} + 2\right) = 8\left(\dfrac{3}{4}\right) \qquad \text{Multiply both sides by 8.}$$

$$8\left(\dfrac{y}{8}\right) + 8(2) = 8\left(\dfrac{3}{4}\right) \qquad \text{Distributive property.}$$

$$
\begin{aligned}
y + 16 &= 6 \qquad &&\text{Simplify.}\\
y + 16 - 16 &= 6 - 16 \qquad &&\text{Subtract 16 from both sides.}\\
y &= -10 \qquad &&\text{Simplify.}
\end{aligned}
$$

Example 3: Solve for m: $\dfrac{m}{4} - \dfrac{m}{7} = 6$

Solution:

$$\dfrac{m}{4} - \dfrac{m}{7} = 6$$

$$28\left(\dfrac{m}{4} - \dfrac{m}{7}\right) = 28(6) \qquad \text{Multiply both sides by the LCD 28.}$$

$$28\left(\dfrac{m}{4}\right) - 28\left(\dfrac{m}{7}\right) = 28(6) \qquad \text{Distributive property.}$$

$$
\begin{aligned}
7m - 4m &= 168 \qquad &&\text{Simplify}\\
3m &= 168 \qquad &&\text{Combine like terms.}\\
\dfrac{3m}{3} &= \dfrac{168}{3} \qquad &&\text{Divide both sides by 3.}\\
m &= 56 \qquad &&\text{Simplify.}
\end{aligned}
$$

4.7 EXERCISES

Solve.

1. $8x = 3$

2. $-6y = 5$

3. $\dfrac{2}{5}a = 10$

4. $-\dfrac{5}{6}m = -\dfrac{7}{12}$

5. $\dfrac{x}{4} + 3 = \dfrac{7}{2}$

6. $\dfrac{y}{8} - \dfrac{8}{16} = 1$

7. $\dfrac{b}{7} - b = -18$

8. $\dfrac{3}{10} - \dfrac{1}{3} = \dfrac{x}{20}$

Add or subtract.

9. $\dfrac{x}{8} - \dfrac{5}{7}$

10. $-\dfrac{4}{11} + \dfrac{y}{6}$

11. $\dfrac{3m}{4} + 9$

12. $\dfrac{8a}{5} - \dfrac{7a}{10}$

Solve. If no equation is given, perform the indicated operation.

13. $\dfrac{5}{9}n = \dfrac{7}{18}$

14. $\dfrac{5}{4} + \dfrac{4}{x} = \dfrac{1}{8}$

15. $\dfrac{11}{12} - \dfrac{5}{6}$

16. $-\dfrac{5}{9}x = \dfrac{5}{18} - \dfrac{7}{18}$

17. $18 - \dfrac{14}{3}$

18. $\dfrac{y}{4} = -3 + y$

19. $\dfrac{6}{7}x = \dfrac{2}{3} - \dfrac{1}{4}$

20. $\dfrac{8}{15}b = -\dfrac{4}{5} + \dfrac{1}{3}$

4.8 OPERATIONS ON MIXED NUMBERS

Example 1: Write $7\dfrac{2}{3}$ as an improper fraction.

Solution: $7\dfrac{2}{3} = \dfrac{3 \cdot 7 + 2}{3}$

$= \dfrac{23}{3}$

Example 2: Write the improper fraction $\dfrac{31}{8}$ as a mixed number.

Solution: Divide 31 by 8.

$$
\begin{array}{r}
3\dfrac{7}{8} \quad \left(\dfrac{\text{remainder}}{\text{divisor}}\right) \\
8\overline{)31} \\
\underline{-24} \\
7
\end{array}
$$

Thus $\dfrac{31}{8} = 3\dfrac{7}{8}$.

Example 3: Multiply: $4\dfrac{1}{4} \cdot \dfrac{16}{7}$

Solution: $4\dfrac{1}{4} \cdot \dfrac{16}{7} = \dfrac{17}{4} \cdot \dfrac{16}{7}$

$= \dfrac{17 \cdot 4 \cdot 4}{4 \cdot 7}$

$= \dfrac{68}{7}$

or $9\dfrac{5}{7}$

Example 4: Subtract: $12\frac{2}{5} - 8\frac{7}{15}$

Solution:
$$12\frac{2}{5} = 12\frac{6}{15} \qquad \text{The LCD of 5 and 15 is 15.}$$

$$\underline{-\ 8\frac{7}{15} = -8\frac{7}{15}}$$

We cannot subtract $\frac{7}{15}$ from $\frac{6}{15}$, so we borrow from the whole number 12.

$$12\frac{6}{15} = 11 + 1\frac{6}{15} = 11 + \frac{21}{15} \quad \text{or} \quad 11\frac{21}{15}$$

$$\begin{array}{r} 11\dfrac{21}{15} \\ -\ 8\dfrac{7}{15} \\ \hline 3\dfrac{14}{15} \end{array}$$
Subtract fractions.
Subtract whole numbers.

4.8 EXERCISES

Write each mixed number as an improper fraction.

1. $9\frac{1}{4}$

2. $2\frac{3}{13}$

3. $12\frac{4}{5}$

Write each improper fraction as a whole number or a mixed number.

4. $\frac{18}{4}$

5. $\frac{57}{3}$

6. $\frac{71}{10}$

Multiply or divide.

7. $3\frac{4}{5} \cdot \frac{1}{8}$

8. $7\frac{5}{6} \cdot 3\frac{2}{3}$

9. $\frac{4}{9} \div 6\frac{3}{4}$

10. $20\frac{3}{8} \cdot 7$

11. $15\frac{2}{7} \div \frac{9}{14}$

12. $12 \div 4\frac{6}{7}$

Add or subtract.

13. $4\frac{3}{14} + 2\frac{5}{7}$

14. $38\frac{3}{5} + 14\frac{7}{10}$

15. $9\frac{7}{8} - 2\frac{1}{4}$

16. $13\frac{1}{6} - 12\frac{2}{3}$

17. $19 - 18\frac{9}{10}$

18. $53\frac{1}{5} + 16\frac{3}{10} + \frac{8}{9}$

Solve.

19. Two packages of sirloin steak weigh $2\frac{3}{5}$ pounds and $3\frac{1}{4}$ pounds. Find their combined weight.

20. Carrie owns $\frac{3}{7}$ of a family bookstore business and John owns $\frac{1}{6}$. Find what part of the business is owned by their only other partner.

CHAPTER 4 PRACTICE TEST

Perform the indicated operations and write the answers in lowest terms.

1. $\frac{5}{5} \div \frac{2}{5}$

2. $-\frac{5}{7} \cdot \frac{9}{5}$

3. $\frac{3x}{8} + \frac{x}{8}$

4. $\frac{1}{10} - \frac{2}{x}$

5. $\frac{x^2 y}{z^3} \cdot \frac{z}{xy^2}$

6. $-\frac{4}{9} \cdot -\frac{16}{30}$

7. $\frac{8b}{11} + \frac{3}{22}$

8. $-\frac{4}{13m} - \frac{6}{13m}$

9. $10x^2 \div \frac{x}{12}$

10. $3\frac{1}{8} \div \frac{5}{24}$

11. $5\frac{1}{8} + 3\frac{1}{2} + 7\frac{3}{4}$

12. $\frac{4x}{7} \cdot \frac{14}{8x^4}$

13. $-\frac{15}{2} \div -\frac{45}{4}$

14. $9\frac{1}{5} \cdot 4\frac{3}{10}$

15. $20 \div 5\frac{3}{4}$

16. $\left(\frac{12}{7} \cdot \frac{21}{6} \right) \div 8$

17. $\frac{5}{6} - \frac{2}{3} + \frac{11}{12}$

Simplify the complex fraction.

18. $\dfrac{\frac{4x}{15}}{\frac{16x^2}{60}}$

19. $\dfrac{4 + \frac{3}{8}}{5 - \frac{1}{4}}$

Solve.

20. $-\frac{7}{8}x = \frac{5}{16}$

21. $\frac{x}{4} + x = -\frac{25}{16}$

22. $\frac{1}{4} + \frac{x}{3} = \frac{5}{6} + \frac{x}{2}$

Evaluate each expression using the given replacement values.

23. $-6x$ for $x = -\dfrac{7}{12}$ 24. xy for $x = \dfrac{2}{3}$ and $y = 4\dfrac{5}{6}$

Solve.

25. Josie has a rope that is $12\dfrac{2}{5}$ feet long. He cuts off a piece $5\dfrac{7}{10}$ feet long. Find the length of the remaining piece of rope.

CHAPTER 4 HELPFUL HINTS AND INSIGHTS

- We cannot divide by 0. This means that the denominator of a fraction may not be 0.

- The number 1 is neither prime nor composite.

- Be careful when all factors of the numerator or denominator are divided out. The result of that numerator or denominator is 1 not 0.

- When the denominator of a fraction contains a variable, such as $\dfrac{8}{5x}$, we assume that the variable does not represent 0.

- Every number has a reciprocal except 0.

- When dividing by a fraction, do not look for common factors to divide out until you rewrite the division as multiplication.

- When working with signed fractions, remember that $-\dfrac{a}{b} = \dfrac{a}{-b} = \dfrac{-a}{b}$.

5.1 INTRODUCTION TO DECIMALS

Example 1: Write the decimal 4.071 in words.

Solution: four   and   seventy-one **thousandths**
                                      ↓
                         4.071
                              ↑_____ thousandths place

Example 2: Write 0.57 as a fraction.

Solution: $0.57 = \dfrac{57}{100}$

         two   **two**
         decimal   zeros
         places

Example 3: Insert <, >, or = to form a true statement.

         38.46   38.458

Solution: 38.46 > 38.458, since 6 > 5.

Example 4: Round 84.352 to the nearest tenth.

Solution: 84.3_5_2 rounded to the nearest tenth is 84.4.

The digit to the right of the tenths place is 5, so we add 1 to the tenths place and drop all the digits to the right of the tenths place.

5.1 EXERCISES

Write each decimal in words.

1. 8.07   2. 142.5   3. 16.341

Write each decimal in standard form.

4. Twelve and four-tenths

5. Fifty-seven and twenty-eight thousandths

6. One hundred eighty-nine and fourteen hundredths

Write each decimal as a fraction or a mixed number, in lowest terms.

7. 0.6   8. 3.72   9. 0.045

10. 14.606   11. 591.44   12. 0.3005

Insert <, >, or = to form a true statement.

13. 0.21   0.22

14. 161.351   161.349

15. 25,000   0.000025

16. 0.13000   0.130

Round the decimal to the given place.

17. 0.49, nearest tenth

18. 2.4631, nearest hundredth

19. 0.5786, nearest thousandth

20. 42,301.89, nearest ten

5.2 ADDING AND SUBTRACTING DECIMALS

Example 1: Add: 42.71 + 3.542.

Solution: Line up the decimal points and add.

$$
\begin{array}{r}
\overset{1}{4}2.71 \\
+\ 3.542 \\
\hline
46.252
\end{array}
$$

Example 2: Add 4.58 + (−6.91).

Solution:

$$
\begin{array}{r}
\overset{8\ 11}{6.9\cancel{1}} \\
-\ 4.58 \\
\hline
2.33
\end{array}
$$
Subtract the absolute values.

4.58 + (−6.91) = −2.33   The answer has the same sign as the number with the larger absolute value.

Example 3: Subtract 2.4 − 0.035.

Solution:

$$
\begin{array}{r}
\overset{\quad 9}{\overset{3\ \cancel{10}\ 10}{2.4\cancel{0}\cancel{0}}} \\
-\ 0.035 \\
\hline
2.365
\end{array}
$$
Two zeros inserted.

Example 4: Is 4.1 a solution of the equation 5.9 = x + 1.8?

Solution:

$5.9 = x + 1.8$
$5.9 = 4.1 + 1.8$ Replace x with 4.1.
$5.9 = 5.9$ True.

Since 5.9 = 5.9 is a true statement, 4.1 is a solution.

5.2 EXERCISES

Perform the indicated operations.

1. $3.6 + 9.2$

2. $14.81 + 16.097$

3. $-7.01 + 6.94$

4. $-39.113 + (-23.04)$

5. $14 - 0.33$

6. $-9.246 - 10.0351$

7. $593.16 - 49.58$

8. Subtract 4.9 from 22.

9. $647.8 - 37.77$

10. $21.647 + (-19.84)$

Find the following when $x = 25$, $y = 3.7$ and $z = 0.106$.

11. $x - z$

12. $x + y + z$

13. $z - y$

See if the given values are solutions to the given equations.

14. $x + 7.3 = 9.8$; $\quad x = 2.6$

15. $51.7 - y = 92$; $\quad y = -40.3$

16. $2.4 - m = m + 0.3$; $\quad m = 1.05$

Simplify by combining like terms.

17. $41.8x + 13.5 - 20.3x - 18.6$

18. $-7.81y - 5.68 - 10.06y + 7.14$

Solve.

19. Gasoline was $1.339 per gallon on one day and $1.359 per gallon the next day. Find by how much the price changed.

20. Find the perimeter of the rectangular garden with dimensions 22.6 feet by 38.4 feet.

5.3 MULTIPLYING DECIMALS AND CIRCUMFERENCE OF A CIRCLE

Example 1: Multiply 13.7×0.51.

Solution:

$$
\begin{array}{rl}
13.7 & \text{1 decimal place} \\
\underline{\times\ 0.51} & \text{2 decimal places} \\
137 & \\
\underline{685\ \ } & \\
6.987 & \text{3 decimal places}
\end{array}
$$

Example 2: Multiply $(-4.3)(0.6)$.

Solution: Recall that the product of a negative number and a positive number is a negative number.

$$(-4.3)(0.6) = -2.58$$

Example 3: Multiply: 580×0.001

Solution: $580 = 580.$

$$580. \times 0.001 \quad = \quad 0.580 \quad \text{or} \quad 0.58$$
3 decimal   3 places
places   to the left

Example 4: Evaluate xy when $x = 9.2$ and $y = 0.53$.

Solution: $xy = (9.2)(0.53)$

$$
\begin{array}{r}
9.2 \\
\times\ 0.53 \\
\hline
276 \\
460 \\
\hline
\end{array}
$$

$= 4.876 \leftarrow \quad 4.876$

5.3 EXERCISES

Multiply.

1. $(0.5)(0.3)$ 2. $(1.68)(0.73)$ 3. $(5.78)(4.1)$ 4. $(1.0006)(4.2)$

5. $(-1.268)(100)$ 6. $(-9.07)(3.5)$ 7. $(-10.206)(-0.01)$ 8. $(561.0704)(1000)$

9. $(2.0056)(7.9)$ 10. $(3.1154)(0.8)$

Evaluate xy using the given replacement values.

11. $x = 35$ and $y = 9.4$ 12. $x = -4.7$ and $y = -6$ 13. $x = -23.1$ and $y = 5.02$

Determine whether the given value is a solution of each given equation.

14. $0.9x = 5.13$; $x = 5.7$ 15. $-0.3x = -10.884$; $x = -36.28$

Find the circumference of each circle. Then use the approximation 3.14 for π and approximate each circumference.

16.
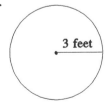
3 feet

17.
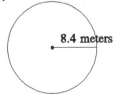
8.4 meters

18.

12 yards

Solve.

19. Billy Rollins is painting a room with paint costing \$15.96 per gallon. If the job requires 2 gallons of paint, find the total cost of the paint.

20. Sandy Okaski is making a dress. The material costs \$9.98 per yard and the dress requires 2.3 yards. Find the total cost.

5.4 DIVIDING DECIMALS

Example 1:   Divide: $3.4\overline{)17.85}$

Solution: Move the decimal points in the divisor and the dividend one place to the right so that the divisor is a whole number.

$$3.4\overline{)17.85} \qquad \text{becomes} \qquad \begin{array}{r} 5.25 \\ 34\overline{)178.50} \\ \underline{170} \\ 85 \\ \underline{68} \\ 170 \\ \underline{170} \\ 0 \end{array}$$

Example 2:   Divide $-4.69 \div 112$.   Round the quotient to the nearest thousandth.

Solution: Recall that a negative number divided by a positive number gives a negative quotient.

$$\begin{array}{r} 0.0418 \approx 0.042 \\ 112\overline{)4.6900} \\ \underline{448} \\ 210 \\ \underline{112} \\ 980 \\ \underline{896} \\ 84 \end{array}$$

If rounding to the nearest thousandth, carry division process out to one more decimal place, the ten-thousandths place.

Thus $-4.69 \div 112 \approx -0.042$.

Example 3:   Divide: $\dfrac{49.6}{1000}$

Solution: $\dfrac{49.6}{1000} \;=\; 0.0496$

$\qquad\quad \uparrow \qquad\qquad \uparrow$

$\qquad$ 3 zeros $\qquad$ 3 decimal places

Example 4:   Is 890 a solution of the equation $\dfrac{x}{2.5} = 365$?

Solution: $\dfrac{x}{2.5} = 365$ Original equation.

$\dfrac{890}{2.5} = 365$ Replace x with 890.

$356 = 365$ False.

Since the statement is false, 890 is not a solution of the equation.

5.4 EXERCISES

Divide.

1. $0.82 \div 4$

2. $-48 \div 0.08$

3. $7.488 \div 0.78$

4. $2.772 \div (-0.33)$

5. $4.812 \div 3$

6. Divide 80 by 0.16

7. Divide 45 by 0.009.

8. $(-3.6) \div (-10,000)$

9. $\dfrac{32.45}{-0.055}$

10. $\dfrac{5.832}{0.081}$

Divide the following. Round the quotients to the nearest hundredth.

11. $\dfrac{85.43}{0.07}$

12. $\dfrac{91.53}{0.13}$

Evaluate the expression when $x = 4.62$, $y = 0.5$, and $z = 3.18$.

13. $z \div y$

14. $x \div 0.002$

Determine whether the given values are solutions of the given equations.

15. $\dfrac{x}{6} = 4.02$; $x = 24.12$

16. $\dfrac{y}{1000} = 0.31$; $y = 3100$

Solve.

17. The area of a rectangle is 80.6 square meters. Its width is 6.5 meters. Find the length of the rectangle.

18. There are approximately 39.37 inches in 1 meter. Determine how many meters there are in 300 inches to the nearest tenth of a meter.

5.5 ESTIMATING AND ORDER OF OPERATIONS

Example 1: Subtract: $84.67 - $19.84. Then estimate the difference to see if the proposed result is reasonable by rounding each decimal to the nearest dollar and then subtracting.

Solution:

Given		Estimate
$84.67	rounds to	$85
$- $19.84	rounds to	$- $20
$64.83		$65

The estimated difference is $65, so $64.83 is reasonable.

Example 2: Divide $311.168 \div 41.6$. Then estimate the quotient to see if the proposed result is reasonable.

Solution: Given   Estimate

$$
\begin{array}{r}
7.48 \\
41.6.\overline{)311.1.68} \\
\underline{291\,2} \\
1996 \\
\underline{1664} \\
3328 \\
\underline{3328} \\
0
\end{array}
\qquad
\begin{array}{r}
7 \\
40\overline{)300} \\
\underline{280}
\end{array}
$$

The estimate is 7, so **7.48** is reasonable.

Example 3: Simplify: $-0.4(9.7 - 2.3)$

Solution: $-0.4(9.7 - 2.3) = -0.4(7.4)$ Inside parentheses first.
                $= -2.96$ Multiply.

Example 4: Evaluate $-3x - 6$ when $x = 9.1$.

Solution: $-3x - 6 = -3(9.1) - 6$ Replace x with 9.1.
                $= -27.3 - 6$ Multiply.
                $= -33.3$ Subtract.

5.5 EXERCISES

Perform each indicated operation. Estimate to see whether each proposed result is reasonable.

1. $3.7 + 9.2 + 4.5$ 2. $16.4 - 8.9$ 3. $(8)(512.3)$

4. $234.25 \div 8.5$ 5. $(-6.3)^2$ 6. $(5.016)(3.42)$

Evaluate each expression when $x = 3.4$, $y = -1.2$ and $z = -0.8$.

7. $-2x + y$ 8. z^2 9. $\dfrac{xy}{z}$

Determine whether the given value is a solution of the given equation.

10. $12x - 4.2 = 3;\ \ x = -0.6$ 11. $5x - 7.4 = 4x + 11.9;\ \ x = 19.3$ 12. $10x + 8.2 = 11x - 9.1;\ \ x = 0.9$

Perform the indicated operations.

13. $(-4.9)^2$ 14. $(5.8 + 0.6)(9.7 - 1.2)$ 15. $\dfrac{0.813 - 4.62}{0.03}$

16. $\dfrac{9 + 0.81}{18}$ 17. $4.3(6.5 - 4.1)$ 18. $2.7 + (0.3)^2$

Solve

19. Estimate the area of a rectangle to the nearest foot if the length is 21.3 feet and the width is 8.7 feet.

20. Use 3.14 for π to approximate the circumference of a circle whose radius is 5 inches.

5.6 FRACTIONS AND DECIMALS

Example 1: Write $\frac{2}{5}$ as a decimal.

Solution:
$$
\begin{array}{r}
0.4 \\
5\overline{)2.0} \\
\underline{2\,0} \\
0
\end{array}
$$

$$\frac{2}{5} = 0.4$$

Example 2: Write 0.16 as a fraction.

Solution: 0.16 is 16 hundredths, so

$$0.16 = \frac{16}{100} = \frac{4}{25}$$

Example 3: Write the numbers in order from smallest to largest.

$$\frac{7}{10}, \quad \frac{5}{8}, \quad 0.65$$

Solution:

Original numbers	$\frac{7}{10}$	$\frac{5}{8}$	0.65
Decimals	0.7	0.625	0.65
Compare in order	3rd	1st	2nd

Then written in order we have $\frac{5}{8}$, 0.65, $\frac{7}{10}$.

5.6 EXERCISES

Write each fraction as a decimal.

1. $\frac{9}{5}$ 2. $\frac{7}{4}$ 3. $\frac{9}{20}$

Write each fraction as a decimal. Round to the nearest hundredth.

4. $\frac{5}{12}$ 5. $\frac{2}{15}$ 6. $\frac{9}{13}$

Write each decimal as a fraction.

7. 0.27   8. 0.425   9. 0.0008

Write each decimal as a mixed number.

10. 19.25   11. 5.74   12. 3.444

Insert <, >, *or* = *to form a true statement.*

13. $\dfrac{4}{9}$ $\dfrac{31}{72}$

14. $\dfrac{21}{23}$ $\dfrac{41}{43}$

15. $\dfrac{71}{12}$ 5.92

16. $\dfrac{538}{19}$ 29.476

Write each number in order from smallest to largest.

17. 0.814, 0.836, 0.83

18. $\dfrac{11}{9}$, 1.22, $\dfrac{13}{8}$

Find the value of each expression.

19. $\dfrac{4}{3} - 5(2.6)$

20. $\dfrac{1}{8}(-10.2 - 21.8)$

5.7 EQUATIONS CONTAINING DECIMALS

Example 1: Solve for x: $x - 3.6 = 12$

Solution:

$$
\begin{aligned}
x - 3.6 &= 12 && \text{Original equation.} \\
x - 3.6 + 3.6 &= 12 + 3.6 && \text{Add 3.6 to both sides.} \\
x &= 15.6 && \text{Simplify.}
\end{aligned}
$$

Example 2: Solve for y: $-5y = 42.5$

Solution:

$$-5y = 42.5 \qquad \text{Original equation.}$$

$$\frac{-5y}{-5} = \frac{42.5}{-5} \qquad \text{Divide both sides by } -5.$$

$$y = -8.5 \qquad \text{Simplify.}$$

Example 3: Solve for x: $4(x - 0.72) = 2x + 3.8$

Solution:

$$
\begin{aligned}
4(x - 0.72) &= 2x + 3.8 && \text{Original equation.} \\
4x - 2.88 &= 2x + 3.8 && \text{Distributive property.} \\
4x - 2.88 + 2.88 &= 2x + 3.8 + 2.88 && \text{Add 2.88 to both sides.} \\
4x &= 2x + 6.68 && \text{Simplify.} \\
4x + (-2x) &= 2x + 6.68 + (-2x) && \text{Add } -2x \text{ to both sides.} \\
2x &= 6.68 && \text{Simplify.}
\end{aligned}
$$

$$\frac{2x}{2} = \frac{6.68}{2} \qquad \text{Divide both sides by 2.}$$

$$x = 3.34 \qquad \text{Simplify.}$$

Example 4: Solve for z: $0.4z + 1.6 = 3.84$

Solution:

$$
\begin{aligned}
0.4z + 1.6 &= 3.84 && \text{Original equation.} \\
100(0.4z + 1.6) &= 100(3.84) && \text{Multiply both sides by 100.} \\
100(0.4z) + 100(1.6) &= 100(3.84) && \text{Distributive property.} \\
40z + 160 &= 384 && \text{Simplify.} \\
40z + 160 + (-160) &= 384 + (-160) && \text{Add } -160 \text{ to both sides.}
\end{aligned}
$$

$$40z = 224 \qquad \text{Simplify.}$$

$$\frac{40z}{40} = \frac{224}{40} \qquad \text{Divide both sides by 40.}$$

$$z = 5.6 \qquad \text{Simplify.}$$

5.7 EXERCISES

Solve the equation.

1. $x + 3.5 = 18.2$

2. $y - 8.9 = 17.3$

3. $6y = 2.58$

4. $0.2m = 15.26$

5. $5x + 11.21 = 6x - 7$

6. $12y - 10.31 = 10y + 1.77$

7. $4(x - 2.9) = 11.2$

8. $7(n + 3.3) = 87.5$

Solve the equation by first multiplying both sides by an appropriate power of 10 so that the equation contains integers only.

9. $0.6x + 0.12 = -0.24$

10. $6x - 12.5 = x$

11. $3.8a + 7 - 1.2a = 22.6$

12. $-0.005x = 29.65$

13. $y + 15.04 = 11.2$

14. $300x - 0.74 = 200x + 0.9$

Solve.

15. $10(x - 9.4) = 25$

16. $12x + 8.6 = 4(2x - 6.1)$

17. $0.9x + 42.1 = x - 57.09$

18. $0.003x - 15 = 0.009$

5.8 SQUARE ROOTS AND THE PYTHAGOREAN THEOREM

Example 1: Find $\sqrt{64}$.

Solution: The square root of 64 is 8 because 8 is positive and $8 \cdot 8 = 64$.

Example 2: Find $\sqrt{\dfrac{1}{100}}$.

Solution: $\sqrt{\dfrac{1}{100}} = \dfrac{1}{10}$ because $\dfrac{1}{10} \cdot \dfrac{1}{10} = \dfrac{1}{100}$.

Example 3: Find the length of the hypotenuse of a right triangle with leg lengths 15 feet and 20 feet.

Solution: Let $a = 15$ and $b = 20$.

$$a^2 + b^2 = c^2$$

$$15^2 + 20^2 = c^2$$

$$225 + 400 = c^2$$

$$625 = c^2$$

$$\sqrt{625} = c \quad \text{or} \quad c = 25$$

The hypotenuse is 25 feet long.

5.8 EXERCISES

1. $\sqrt{196}$ 2. $\sqrt{0}$ 3. $\sqrt{225}$ 4. $\sqrt{\dfrac{16}{81}}$ 5. $\sqrt{\dfrac{1}{121}}$ 6. $\sqrt{\dfrac{144}{49}}$

Use a calculator to approximate the square root. Round the square root to the nearest thousandth.

7. $\sqrt{11}$ 8. $\sqrt{21}$ 9. $\sqrt{37}$ 10. $\sqrt{84}$ 11. $\sqrt{92}$ 12. $\sqrt{143}$

Find the length of the hypotenuse of each right triangle with given leg lengths. If necessary, approximate the length to the nearest thousandth.

13. leg = 18, leg = 24

14. leg = 7, leg 9

15. leg = 21, leg = 30

16. leg = 15, leg = 36

Solve.

17. Find the length of the diagonal of the rectangle to the nearest hundredth of a foot.

20 feet

40 feet

18. Tammy Henderson has a rectangular frame with dimensions 24 inches by 14 inches. Find the length of the diagonal of the frame to the nearest hundredth of an inch.

CHAPTER 5 PRACTICE TEST

Write the decimal as indicated.

1. 84.17, in words

2. Two hundred sixty-eight and seven tenths, in standard form

Perform each indicated operation. Round the result to the nearest thousandth if necessary.

3. 5.716 + 2.89 + 12.398

4. −61.87 − 4.76

5. 8.39 − 27.64

6. (13.5)(5.26)

7. (−0.03834) ÷ (−0.71)

Round the decimal to the indicated place value.

8. 47.9746, nearest hundredth

9. 0.1478, nearest thousandth

Insert <, >, or = to make a true statement.

10. 51.0807   51.078

11. $\dfrac{3}{7}$ 0.429

Write the decimal as a fraction or a mixed number.

12. 0.625

13. 18.97

Write the fraction as a decimal.

14. $\dfrac{17}{85}$

15. $\dfrac{19}{40}$

Simplify.

16. $(-0.4)^2 + 2.68$

17. $\dfrac{0.56 + 2.34}{-0.2}$

18. $12.1x - 6.4 - 9.8x - 7.6$

Find each square root and simplify. Round the square root to the nearest thousandth if necessary.

19. $\sqrt{25}$

20. $\sqrt{\dfrac{121}{36}}$

21. $\sqrt{175}$

Solve.

22. Approximate to the nearest hundredth of an inch the length of the missing side of a right triangle with legs of 7 inches each.

23. Approximate to the nearest hundredth of a meter the length of the missing side of a right triangle with legs of 9 meters and 11 meters.

24. $0.6x + 5.7 = 0.3$

25. $12(x + 1.6) = 10x - 8.4$

CHAPTER 5 HELPFUL HINTS AND INSIGHTS

- When writing a decimal from words to decimal notation, make sure the last digit is in the correct place by inserting 0 s if necessary.

- For any decimal inserting 0 s to the right of the decimal point after the last digit does not change the value of the number.

- When a whole number is written as a decimal, the decimal point is placed to the right of the ones digit.

- Zeros may be inserted to the right of the decimal point after the last digit to help line up place values when adding decimals.

- $\sqrt{43}$ is **approximately** 6.557. This means that, if we multiply 6.557 by 6.557, the product is **close** to 43.

6.1 RATIOS

Example 1: Write the ratio of 9 to 13 using fractional notation.

Solution: $\dfrac{9}{13}$ Remember order is important.

Example 2: Write the ratio of 25 cents to 85 cents as a fraction in simplest form.

Solution: $\dfrac{25 \text{ cents}}{85 \text{ cents}} = \dfrac{25}{85} = \dfrac{5 \cdot 5}{17 \cdot 5} = \dfrac{5}{17}$

6.1 EXERCISES

Write each ratio using fractional notation. Do not simplify.

1. 23 to 31

2. 100 to 53

3. $\dfrac{1}{3}$ to 8

4. $5\dfrac{4}{5}$ to $2\dfrac{1}{10}$

Write each ratio as a ratio of whole numbers using fractional notation. Write the fraction in simplest form.

5. 18 to 57

6. 75 to 375

7. 22 meters to 162 meters

8. 80 feet to 324 feet

9. $116 to $20

10. 84 miles to 105 miles

11. 7.2 to 103

12. 3.24 to 5.28

Solve.

In Mrs. Roberts' Calculus class there were 18 men and 14 women present.

13. Find the ratio of women to men.

14. Find the ratio of men to total people present.

Use the given figure to find the ratio described in each problem.

6 inches

20 inches

15. Find the ratio of the width to the length of the rectangle.

16. Find the ratio of the length to the perimeter of the rectangle.

6.2 RATES

Example 1: Write the rate as a fraction in simplest form: 21 flowers every 6 feet

Solution: $\dfrac{21 \text{ flowers}}{6 \text{ feet}} = \dfrac{7 \text{ flowers}}{2 \text{ feet}}$

Example 2: Write "$950 every 5 months" as a unit rate.

Solution: $\dfrac{950 \text{ dollars}}{5 \text{ months}}$

$$5\overline{)950} \quad \begin{array}{c} 190 \end{array}$$

$= \dfrac{190 \text{ dollars}}{1 \text{ month}}$

or 190 dollars/month

Example 3: Approximate each unit price to decide which is the better buy:

$1.88 for 8 poptarts or $1.49 for 6 poptarts.

Solution: $\dfrac{\$1.88}{8 \text{ poptarts}} \approx 0.24 \text{ per poptart}$

$\dfrac{\$1.49}{6 \text{ poptarts}} \approx 0.25 \text{ per poptart}$

Thus the 8 poptart box is the better deal.

6.2 EXERCISES

Write each rate as a fraction in simplest form.

1. 10 lightpoles every 2 miles

2. 36 computers for 78 faculty members

3. 9 defective lightbulbs out of every 990 lightbulbs

4. 20 pear trees every 220 feet

5. 12 rooms for 116 people

6. 40 boxes for 192 cookies

Write each of the following as a unit rate.

7. 540 calories in a 5-ounce serving

8. 486 miles in 9 hours

9. $2.24 for 7 pears

10. 432 miles for 18 gallons of gas

11. $7,000,000 for 14 lottery winners

12. 1050 students for 35 teachers

Find each unit rate and decide which is the better buy. Round to the nearest cent.

13. Frozen grape juice: $1.82 for 16 ounces
     $1.12 for 9 ounces

14. Donuts: $1.00 for 3 donuts
     $3.60 for 12 donuts

15. Frozen shrimp: $7.99 for 20 ounces
     $4.99 for 12 ounces

6.3 PROPORTIONS

Example 1: Is $\dfrac{4}{5} = \dfrac{8}{9}$ a true proportion?

Solution: $\dfrac{4}{5} = \dfrac{8}{9}$

$(4)(9) = (5)(8)$
$36 = 40$
False

Since the cross products are not equal, the proportion is not a true statement.

Example 2: Solve: $\dfrac{8}{3} = \dfrac{x}{27}$

Solution: $\dfrac{8}{3} = \dfrac{x}{27}$

$8 \cdot 27 = 3 \cdot x$ Cross multiply.
$216 = 3x$ Multiply.

$\dfrac{216}{3} = \dfrac{3x}{3}$ Divide both sides by 3.

$72 = x$

Example 3: Solve for y: $\dfrac{2.7}{3.4} = \dfrac{0.8}{y}$. Round the solution to the nearest hundredth.

Solution: $\dfrac{2.7}{3.4} = \dfrac{0.8}{y}$

$2.7 \cdot y = 3.4 \cdot 0.8$ Cross multiply.
$2.7y = 2.72$

$\dfrac{2.7y}{2.7} = \dfrac{2.72}{2.7}$ Divide both sides by 2.7

$y = 1.01$

6.3 EXERCISES

Write the sentence as a proportion.

1. 20 books is to 4 students as 30 books is to 6 students.

2. $1\dfrac{1}{2}$ cups of flour is to 10 crepes as $3\dfrac{3}{4}$ cups of flour is to 25 crepes.

3. 18 errors is to 12 pages as 3 errors is to 2 pages.

4. 36 inches is to 3 feet as 144 inches is to 12 feet.

Determine whether the proportion is a true proportion.

5. $\dfrac{16}{12} = \dfrac{40}{32}$

6. $\dfrac{40}{60} = \dfrac{1200}{1800}$

7. $\dfrac{0.9}{0.2} = \dfrac{4.5}{1.0}$

8. $\dfrac{2.8}{1.7} = \dfrac{14}{9}$

9. $\dfrac{\frac{3}{5}}{\frac{7}{10}} = \dfrac{\frac{2}{5}}{\frac{7}{15}}$

10. $\dfrac{4\frac{1}{8}}{\frac{7}{3}} = \dfrac{24\frac{3}{4}}{14}$

Solve the proportion for the variable. Approximate the solution when indicated.

11. $\dfrac{x}{4} = \dfrac{110}{44}$

12. $\dfrac{7}{y} = \dfrac{84}{132}$

13. $\dfrac{25}{80} = \dfrac{z}{16}$

14. $\dfrac{7}{8} = \dfrac{98}{n}$

15. $\dfrac{\frac{3}{7}}{21} = \dfrac{x}{7}$

16. $\dfrac{9.4}{3.2} = \dfrac{4.7}{y}$

17. $\dfrac{\frac{8}{9}}{\frac{26}{27}} = \dfrac{2\frac{2}{3}}{z}$

18. $\dfrac{0.6}{y} = \dfrac{12}{400}$

19. $\dfrac{x}{6.12} = \dfrac{0.91}{0.07}$. Round to the nearest tenth.

20. $\dfrac{2036}{5694} = \dfrac{3122}{y}$. Round to the nearest hundredth.

6.4 PROPORTIONS AND PROBLEM SOLVING

Example 1: On a AAA map of Colorado Springs, 7.5 miles corresponds to 3 inches. How many miles correspond to 8 inches?

Solution: Let x = the number of miles represented by 8 inches

$$\dfrac{7.5 \text{ miles}}{3 \text{ inches}} = \dfrac{x \text{ miles}}{8 \text{ inches}}$$

$\begin{aligned} 7.5 \cdot 8 &= 3x \\ 60 &= 3x \end{aligned}$ Cross multiply.

$\dfrac{60}{3} = \dfrac{3x}{3}$ Divide both sides by 3.

$20 = x$

20 miles corresponds to 8 inches.

Example 2:   Six boxes of cereal cost $17.10. How much should 11 boxes cost?

Solution: Let x = cost of 11 boxes

$$\frac{6 \text{ boxes}}{\$17.10} = \frac{11 \text{ boxes}}{x \text{ dollars}}$$

$6 \cdot x = 17.10 \cdot 11$ Cross multiply.
$6x = 188.1$

$$\frac{6x}{6} = \frac{188.1}{6}$$ Divide both sides by 6.

$x = 31.35$

11 boxes should cost $31.35.

6.4 EXERCISES

Solve.

The ratio of a quarterback's completed passes to attempted passes is 5 to 12.

1. If he attempted 36 passes, find how many passes he completed.

2. If he completed 25 passes, find how many passes he attempted.

On an architect's blueprint, 1 inch corresponds to 6 feet.

3. Find the length of a wall represented by a line $3\frac{1}{3}$ inches long on the blueprint.

4. If the exterior wall is 48 feet long, find how long the blueprint measurement should be.

A bag of fertilizer covers 5000 square feet of lawn.

5. Find how many bags of fertilizer should be purchased to cover a rectangular lawn 320 feet by 210 feet.

6. Find how many bags of fertilizer should be purchased to cover a square lawn 220 feet on each side.

Yearly homeowner property taxes are figured at a rate of $1.65 tax for every $100 of house value.

7. If Lynda a homeowner pays $3465 in property taxes, find the value of her home.

8. Find the property taxes on a home valued at $158,000.

An Orioles baseball player makes 4 hits in every 9 times at bat.

9. If this Orioles player comes up to bat 36 times, find how many hits he would be expected to make.

10. At this rate, if he made 12 hits, find how many times he batted.

A survey reveals that 5 out of 7 people prefer cake to pie.

11. In a room of 84 people, how many people are likely to prefer cake?

12. In a college class of 35 students, find how many students are likely to prefer pie.

A Chevrolet Impala averages 400 miles on a 16-gallon tank of gas.

13. If Nancy Varner buys 6 gallons of gas, for her Chevrolet Impala determine how far she can go on that amount. Round to the nearest mile.

14. Find how many gallons of gas Paul Cantwell can expect to burn on a 1560-mile trip in a Chevrolet Impala. Round to the nearest gallon.

6.5 SIMILAR TRIANGLES AND PROBLEM SOLVING

Example 1: If the following two triangles are similar, find the unknown length x.

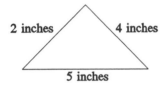

Solution: Let x = length of unknown side

$$\frac{5}{x} = \frac{2}{1}$$

$5 \cdot 1 = x \cdot 2$
$\quad 5 = 2x$ Simplify.

$$\frac{5}{2} = \frac{2x}{2}$$ Divide both sides by 2.

$\quad 2.5 = x$ Simplify.

Thus, x is 2.5 inches.

Example 2: If a 28-foot pole casts a 36-foot shadow, find the length of the shadow cast by a 48-foot pole. Round to the nearest tenth.

Solution:

$$\frac{28}{48} = \frac{36}{x}$$

$$28 \cdot x = 48 \cdot 36$$
$$28x = 1728$$

$$\frac{28x}{28} = \frac{1728}{28}$$

$$x = 61.7$$

The shadow is 61.7 feet.

6.5 EXERCISES

Find the ratio of the corresponding sides of the similar triangles.

1.

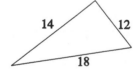

2.

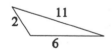

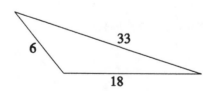

Given that the pairs of triangles are similar, find the length of the side labeled x.

3.

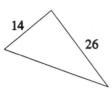

4.

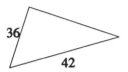   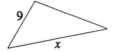

5.

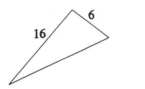   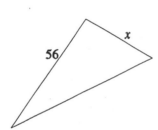

6.

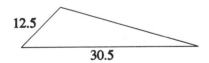   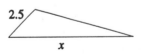

7.

8.

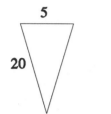

Solve.

9. If a 40-foot tree casts a 22-foot shadow, find the length of the shadow cast by a 52-foot tree.

10. Lawrence, a 6-foot-tall forest ranger, needs to know the height of a tree. He notices that when the shadow of the tree is 54 feet long his shadow is 5 feet long. Find the height of the tree.

CHAPTER 6 PRACTICE TEST

Write each ratio using fractional notation. Do not simplify.

1. Write the ratio as a fraction in lowest terms: 381 roses to 574 roses.

2. On Wednesday the price of ABC stock was $9\frac{1}{4}$ dollars, but the stock rose to $12\frac{3}{8}$ dollars on Thursday. Find the ratio of Wednesday's price to Thursday's price.

Write each ratio as a ratio of whole numbers using fractional notation. Write the fraction in simplest form.

3. 90 to 27

4. 75 to 125

5. $130 to $80

6. 36 feet to 150 feet

Write each rate as a fraction in simplest form.

7. 5 houses every 3 miles

8. 12 fax machines for 92 faculty members

Write each phrase as a unit rate.

9. 570 kilometers in 6 hours

10. 18 inches of snow in 30 days

Compare the unit rates and decide which is the better buy.

11. Taco shells: 12 for $1.49 or 18 for $1.79

12. Potatoes: 3 pounds for $0.89 or 10 pounds for $1.99

Write the sentence as a proportion.

13. 6 cups of water is to 2 bottles as 24 cups of water is to 8 bottles.

14. 81 feet is to 27 yards as 36 feet is to 12 yards.

Determine whether the proportion is a true proportion.

15. $\dfrac{15}{75} = \dfrac{90}{450}$

16. $\dfrac{\frac{1}{3}}{\frac{2}{7}} = \dfrac{\frac{20}{4}}{\frac{9}{2}}$

Solve the proportion for the variable.

17. $\dfrac{x}{4} = \dfrac{35}{20}$

18. $\dfrac{\frac{18}{10}}{\frac{7}{11}} = \dfrac{x}{\frac{5}{9}}$

19. $\dfrac{3.6}{4} = \dfrac{9}{y}$

Solve.

On a map 1 inch corresponds to 50 miles.

20. Find how far apart 2 cities are if their corresponding points on the map are 17 inches apart.

21. Jane Miller's hometown is 45 miles from Betsy Rio's hometown. How far apart will they be on the map?

22. Given that the following triangles are similar, find the missing length.

23. If 6.4 ft of material is used to manufacture one tablecloth, how much material is needed for 5 tablecloths?

24. Kelly Dawson is trying to estimate the height of a tree. She estimates the length of her shadow to be 3½ feet long and the length of the tree's shadow to be 32 feet long. Find the height of the tree if Kelly is 5¼ feet tall.

CHAPTER 6 HELPFUL HINTS AND INSIGHTS

- When comparing quantities with different units, write the units as part of the comparison.

- When writing proportions to solve problems, we will place the same units in the numerators and the same units in the denominators.

7.1 PERCENTS, DECIMALS, AND FRACTIONS

Example 1: Write 68% as a decimal.

Solution: 68% = 0.68   Move the decimal point two places to the left and drop the % symbol.

Example 2: Write 0.765 as a percent.

Solution: 0.765 = 76.5%   Move the decimal point two places to the right and attach the % symbol.

Example 3: Write 2.8% as a fraction in lowest terms.

Solution: $2.8\% = \dfrac{2.8}{100}$

$\qquad = \dfrac{2.8}{100} \cdot \dfrac{10}{10}$ Multiply numerator and denominator by 10.

$\qquad = \dfrac{28}{1000}$

$\qquad = \dfrac{7}{250}$ Reduce.

Example 4: Write $\dfrac{7}{10}$ as a percent.

Solution: $\dfrac{7}{10} = \dfrac{7}{10} \cdot 100\%$

$\qquad = \dfrac{700}{10}\%$

$\qquad = 70\%$

7.1 EXERCISES

Write each percent as a decimal.

1. 52%   2. 5%   3. 81.6%   4. 190%

Write each decimal as a percent.

5. 0.99   6. 0.037   7. 2.12   8. 14

Write each percent as a fraction in simplest form.

9. 16%   10. 118%   11. 3.2%   12. 0.6%

Write each fraction or mixed number as a percent.

13. $\dfrac{3}{8}$ 14. $\dfrac{7}{20}$ 15. $1\dfrac{9}{10}$ 16. $\dfrac{9}{40}$

17. If 25 students are enrolled in a class and the attendance today is 100%, determine how many students attended.

18. The Talbott family saves 0.05 of their take-home pay. Write 0.05 as a percent.

19. A carpet salesman receives a commission of 3.5% of his sales. Write 3.5% as a decimal.

20. A dress is on sale for 3/4 of the original price. Write 3/4 as a percent.

7.2 SOLVING PERCENT PROBLEMS WITH EQUATIONS

Example 1: What is 16% of 80?

Solution: What is 16% of 80?
$$\downarrow \quad \downarrow \quad \downarrow \quad \downarrow \quad \downarrow$$
$$x \; = \; 16\% \; \cdot \; 80$$

$x = 0.16 \cdot 80$ Write 16% as 0.16.
$x = 12.8$

12.8 is 16% of 80.

Example 2: 3.2% of what number is 2.24?

Solution: $3.2\% \cdot x = 2.24$ Write as an equation.
$0.032x = 2.24$ Write 3.2% as a decimal.

$$\frac{0.032x}{0.032} = \frac{2.24}{0.032}$$ Divide both sides by 0.032.

$$x = 70$$

3.2% of 70 is 2.24.

Example 3: 0.12 is what percent of 60?

Solution: $0.12 = x \cdot 60$ Write as an equation.
$0.12 = 60x$

$$\frac{0.12}{60} = \frac{60x}{60}$$ Divide both sides by 60.

$0.002 = x$
$0.2\% = x$ Write 0.002 as a percent.

0.12 is 0.2% of 60.

7.2 EXERCISES

Translate each to an equation. Do not solve.

1. 35% of 80 is what number?

2. What percent of 12 is 8?

3. 6.2 is 29% of what number?

4. 102% of 50 is what?

Solve. If necessary, round to the nearest hundredth.

5. 5% of 40 is what number?

6. What is 18% of 70?

7. 40 is 20% of what number?

8. 0.36 is 52% of what number?

9. 9 is what percent of 36?

10. 8.25 is what percent of 82.5?

11. 0.8 is 20% of what number?

12. 70 is 100% of what number?

13. 14.2 is 8¼% of what number?

14. 520 is what percent of 65?

15. 21.3 is what percent of 100?

16. 90% of what number is 525.6?

7.3 SOLVING PERCENT PROBLEMS WITH PROPORTIONS

Example 1: Translate to a proportion:

8% of what number is 16?

Solution:

$$\underset{\substack{\downarrow \\ \text{percent}}}{8\%} \text{ of } \underset{\substack{\downarrow \\ \text{base}}}{\text{what number}} \text{ is } \underset{\substack{\downarrow \\ \text{amount}}}{16?}$$

$$\begin{array}{l} \text{amount} \to \underline{16} = \underline{8} \leftarrow \text{percent} \\ \text{base} \to b = 100 \end{array}$$

Example 2: What number is 22% of 80?

Solution:

$$\underset{\substack{\downarrow \\ \text{amount}}}{\text{What number}} \text{ is } \underset{\substack{\downarrow \\ \text{percent}}}{22\%} \text{ of } \underset{\substack{\downarrow \\ \text{base}}}{80?}$$

$$\frac{a}{80} = \frac{22}{100}$$

$$a \cdot 100 = 80 \cdot 22$$
$$100a = 1760$$

$$\frac{100a}{100} = \frac{1760}{100}$$

$$a = 17.6$$

17.6 is 22% of 80.

Example 3:   What percent of 90 is 36?

Solution: What percent of 90 is 36?

$$\underbrace{\text{What percent}}_{\text{percent}} \text{ of } \underset{\text{base}}{90} \text{ is } \underset{\text{amount}}{36}?$$

$$\frac{36}{90} = \frac{p}{100}$$

$$36 \cdot 100 = 90 \cdot p$$
$$3600 = 90p$$

$$\frac{3600}{90} = \frac{90p}{90}$$

$$40 = p$$

Then, 40% of 90 is 36.

7.3 EXERCISES

Translate each to a proportion. Do not solve.

1. What percent of 51 is 17?

2. 9% of what number is 10?

3. 11 is 37% of what number?

4. 70% of 112 is what number?

Solve. If necessary, round to the nearest hundredth.

5. 15% of 70 is what number?

6. What is 12% of 50?

7. 60 is 8% of what number?

8. 0.7 is 80% of what number?

9. 60 is what percent of 110?

10. 12 is what percent of 112?

11. 120 is 95% of what number?

12. 45 is 100% of what number?

13. 31.25 is $6\frac{1}{4}$% of what number?

14. 200 is what percent of 50?

15. 27.6 is what percent of 100?

16. 110% of what number is 60?

7.4 APPLICATIONS OF PERCENT

Example 1: Dave Scango counted 5 students absent in his Prealgebra class on a particular day. If this is 20% of the students in his class, how many students are in Dave's Prealgebra class?

Solution: *Method* 1.

In words: 5 is 20% of what number?

Translate: $5 = 20\% \cdot x$

$5 = 0.20x$

$\dfrac{5}{0.20} = \dfrac{0.20x}{0.20}$

$25 = x$

There are 25 students in Dave's class.

Method 2.

In words: 5 is 20% of what number?

amount percent base

Translate: $\dfrac{5}{b} = \dfrac{20}{100}$

$5 \cdot 100 = b \cdot 20$

$500 = 20b$

$\dfrac{500}{20} = \dfrac{20b}{20}$

$25 = b$

There are 25 students in Dave's class.

Example 2:   The number of students taking Prealgebra at the local community college increased from 80 to 95 in one year. What is the percent increase? Round to the nearest whole percent.

Solution: amount of increase = 95 − 80 = 15

$$\text{percent of increase} = \frac{\text{amount of increase}}{\text{original amount}}$$

$$= \frac{15}{80} \approx 0.19 = 19\%$$

The number of students increased by about 19%.

Example 3:   In an effort to increase sales the Flower Gallery decreased the price of their Spring Bright Bouquet from $65 to $55. What was the percent decrease in the price? Round to the nearest whole percent.

Solution: amount of decrease = 65 − 55 = 10

$$\text{percent of decrease} = \frac{\text{amount of decrease}}{\text{original amount}}$$

$$= \frac{10}{65} \approx 0.15 = 15\%.$$

7.4 EXERCISES

Solve. If necessary, round to the nearest hundredth.

1. A 4.5% sales tax is charged on a $1600 copier. 4.5% of $1600 is what number?

2. Tim's parents gave him $500 towards college expenses. He spent $325 of this money on textbooks. What percent of the $500 was spent on textbooks?

3. The social security taxes an employee pays are 15.02% of total wages. Find the amount of social security tax if wages are $1250.

4. Last year, Dave bought a share of stock for $90. He was paid a dividend of $7.38. Determine what percent of the stock price is the dividend.

Solve. Round all percents to the nearest tenth, if necessary.

5. When switching brands of gasoline, Kim Menzies increased her car's rate of miles per gallon from 22.4 to 26.2. Find the percent increase.

6. The number of employees at Klines increased from 6230 to 7110 over the past year. Find the percent increase.

7. Roberta White went on a diet and decreased her normal calorie intake of 2100 back to 1600. Find the percent decrease.

8. Due to a slump in the economy a house valued at $208,000 fell to a value of $178,000. Find the percent decrease.

9. Gasoline increased in price per gallon from $1.16 to $1.32. Find the percent increase.

10. The number of plumbers at Reliance Plumbing decreased from 42 to 36 in one year. Find the percent decrease.

7.5 PERCENT AND PROBLEM SOLVING: SALES TAX, COMMISSION, AND DISCOUNT

Example 1: Find the sales tax and the total price on a purchase of $65.25 in a town where the sales tax rate is 6.5%.

Solution:

Sales tax = tax rate · purchase price
= 6.5% · $65.25
= (0.065)($65.25)   Write 6.5% as a decimal.
≈ $4.24   Rounded to the nearest cent.

Total price = purchase price + sales tax
= $65.25 + $4.24
= $69.49

The sales tax on $65.25 is $4.24 and the total price is $69.49.

Example 2: Jocelyn works at Dress Divine. She earned $96 dollars for selling $1200 worth of merchandise. Find the commission rate.

Solution:

commission = commission rate · sales

$\downarrow$ $\downarrow$ $\downarrow$

96 = r · 1200

$$\frac{96}{1200} = \frac{1200r}{1200}$$

$0.08 = r$ Simplify.
$8\% = r$ Write 0.08 as a percent.

The commission rate is 8%.

Example 3: Woodie's increases the price of a $30 bottle of men's cologne by 5%. What is the increase in the price, and what is the new price of the cologne?

Solution:

increase = 5% · $30
= (0.05)(30)   Write 5% as a decimal.
= 1.5   Multiply.

The increase in price is $1.50.

new price = original price + increase
= $30 + $1.50
= $31.50

The bottle of men's cologne now costs $31.50.

Example 4: The price of a $300 VCR is reduced 30%. What is the decrease and what is the new price?

Solution: decrease = 30% · $300

                = (0.30)($300)   Write 30% as a decimal.

                = $90   Multiply.

The decrease in price is $90.

    new price = original price − decrease

               = $300 − $90

               = $210

The new reduced price is $210.

7.5 EXERCISES

Solve.

1. What is the sales tax on a dress priced at $56 if the sales tax rate is 8%?

2. If the sales tax rate is 5%, find the sales tax on a VCR priced at $250.

3. A CD player has a price of $350. What is the sales tax if the sales tax rate is 4.5%?

4. An area rug is priced at $1200. The sales tax rate is 7.5%. Find the total price.

5. A television is priced at $799. The sales tax rate is 6%. Find the total price.

6. Ms. Rath bought a sweater for $45, a dress for $75, and a blazer for $120. Find the total price she paid, given a sales tax rate of 5.5%.

7. How much commission will Tom Bailey make on the sale of a $245,000 house if he receives 1.2% of the selling price?

8. A salesperson earned a commission of $2675 for selling $29,722 worth of computer products. Find the commission rate.

9. Andrea Barna sold $18,650 worth of sports equipment this week. Find her commission for the week, if she receives a commission rate of 5.5%.

10. The price of a $800 chair is on sale at 20% off. Find the discount and the sale price.

11. The price of a $125 telephone is on sale at 12% off. Find the discount and the sale price.

12. The price of a $25,000 automobile is on sale at 3% off. Find the discount and the sale price.

7.6 PERCENT AND PROBLEM SOLVING: INTEREST

Example 1: Find the simple interest after 4 years on $600 at an interest rate of 5%.

Solution: In this example, P = \$600, R = 5%,   and T = 4 years.

$I = P \cdot R \cdot T$ Simple interest formula.
$ = \$600 \cdot 5\% \cdot 4$ Replace the variables by their values.
$ = \$600 \cdot 0.05 \cdot 4$ Write 5% as a decimal.
$ = \$120$ Multiply.

The simple interest is \$120.

Example 2:   \$9000 is invested at 7% compounded quarterly for 5 years. Find the total amount at the end of 5 years.

Solution: Look in Appendix F of the textbook for the compound interest factor. The compound interest factor for 5 years at 7% in the compounded quarterly section is 1.41478.

Total amount = original principal $\cdot$ compound interest factor
$ = \$9000(1.41478)$
$ = \$12733.02$

The total amount in the account at the end of 5 years is \$12,733.02.

Example 3:   Find the monthly payment for a \$3000 loan for 4 years. The interest in the 4-year loan is \$1720.56.

Solution:
Principal	+	interest	=	Total amount borrowed
\$3000	+	\$1720.56	=	\$4720.56

The number of monthly payments is (4 years)(12 payments/year) = 48 payments.

$$\text{Monthly payment} = \frac{\text{principal + interest}}{\text{total number of payments}}$$

$$= \frac{\$4720.56}{48}$$

$$\approx \$98.35$$

7.6 EXERCISES

Principal	Rate	Time		Principal	Rate	Time
1. \$750	4%	5 years		2. \$8000	5.5%	6 years
3. \$400	14%	2½ years		4. \$650	16.5%	20 months
5. \$200	13%	3 months		6. \$3000	7%	6¼ years

Solve.

7. A company borrows \$75,000 for 3 years at a simple interest of 11.5%. Find the simple interest.

8. A money market fund advertises a simple interest rate of 8.5%. Find the simple interest on \$6250 for 1½ years.

Find the total amount in each compound interest account.

9. $9500 is compounded annually at a rate of 12% for 10 years.

10. $3025 is compounded semiannually at a rate of 17% for 5 years.

11. $12,000 is compounded quarterly at a rate of 8% for 15 years.

12. $22,000 is compounded daily at a rate of 6% for 20 years.

13. $950 is compounded annually at a rate of 5% for 5 years.

14. $1475 is compounded quarterly at a rate of 16% for 10 years.

Solve.

15. $30,000 is borrowed for 5 years. If the interest on the loan is $14,578.42, find the monthly payment.

16. $128,000 is borrowed for 30 years. If the interest on the loan is $232,000, find the monthly payment.

CHAPTER 7 PRACTICE TEST

1. Write 0.127 as a percent.

2. Write 0.3% as a decimal.

3. Write 140% as a fraction.

4. Write $\dfrac{19}{20}$ as a percent.

5. What is 35% of 90?

6. 0.4% of what number is 10.8?

7. 507 is what percent of 845?

8. Write an equation for the statement, "39.2 is 18% of what number?"

9. The membership in the Math Club increased 10.4%. Write this percent as a fraction.

10. Write a proportion for the statement, "29 is what percent of 185?"

Solve. Round all dollar amounts to the nearest cent and all percents to the nearest tenth of a percent.

11. An alloy is 16% copper. How much copper is contained in 180 pounds of this alloy?

12. A fruit grower in Florida estimates that 25% of his potential crop, or $14,575 has been lost to a hard freeze. Find the total value of his potential crop.

13. A real estate agent received a commission rate of 1.6% on the sale of a house for $185,000. Find the commission.

14. If the local sales tax rate is 4.5%, find the total amount charged for a television priced at $650.

15. Jenny borrowed $600 from a bank at 15.5% for 9 months. Find the total simple interest due the bank at the end of the 9-month period.

16. Find the simple interest earned on $4000 saved for 5½ years at a interest rate of 7.75%.

17. The price of a gallon of milk increased 16%. Originally it cost $2.15 per gallon. Find the new price.

18. The number of customers at Jose's Hair Salon decreased from 580 to 510 over the last year. Find the percent decrease.

19. $5 is invested at a local bank at 8% compounded quarterly. Find the total amount after 20 years.

20. $5260 is compounded semiannually at 6%. Find the total amount in the account after 10 years.

21. The price of milk at the local supermarket has increased from $2.50 per gallon to $3.09 per gallon over the last six months. Find the percent increase.

22. $8800 is borrowed for 2½ years. If the interest on the loan is $620.35, find the monthly payment.

23. Diane Nelson, a real estate broker, sold a house for $200,000 last week. If her commission is 1.8% of the selling price of the home, find the amount of her commission.

CHAPTER 7 HELPFUL HINTS AND INSIGHTS

- Remember that, unless stated otherwise, the interest rate given is **per year**. A time period given other than years must be converted to years.

- Before solving an equation that has been translated from words, be sure to convert given percents to decimal form.

8.1 READING CIRCLE GRAPHS

Example 1: The circle graph below is a result of surveying 500 college students. They were asked what type of movie they preferred.

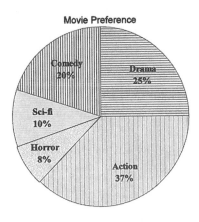

Movie Preference

a. What type of movie is preferred most by the students in this survey?

b. Find the number of students out of the survey who prefer comedy.

Solution: a. Look for the largest sector of the circle, it will be labeled with the largest percent. It is Action.

b. amount = percent · base
amount = 20% · 500
amount = (0.20)(500)
amount = 100

100 students prefer comedy.

Example 2: The following table shows the percent of fish that were stocked in Lake Aqua on Tuesday. Draw a circle graph showing the data.

Type of Fish	Percent
Rainbow Trout	50%
Brown Trout	10%
Large Mouth Bass	25%
Small Mouth Bass	15%

Solution: Find the number of degrees in each sector representing each type of fish.

Sector	Degrees in Each Sector
Rainbow Trout	50% x 360° = 180°
Brown Trout	10% x 360° = 36°
Large Mouth Bass	25% x 360° = 90°
Small Mouth Bass	15% x 360° = 54°

Now using a protractor mark off 4 sectors with the degrees shown in the chart above.

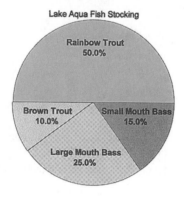

8.1 EXERCISES

Use the circle graph from Example 1 to do Exercises 1-4.

1. How many students prefer horror movies?

2. How many students prefer either sci-fi or action movies?

3. Find the ratio of students preferring action movies to total students.

4. Find the ratio of students preferring comedy to students preferring drama.

The circle graph below shows how many students participated in an activity at Camp Takabreak for a particular session. Use this graph for Exercises 5-10.

5. Which category had the most student participation?

6. Which category had the least student participation?

7. What percent of the students went fishing?

8. Find the ratio of students that went canoeing to the students that played volleyball.

9. Find the ratio of students participating in either swimming or softball to total students.

10. Which activity had the second-largest participation?

11. Draw a circle graph to represent the following **data**.

Type of Glassware in a Collection	
Fenton	40%
Smith	6%
Mosser	12%
Victorian	18%
Westmoreland	24%

8.2 READING PICTOGRAPHS, BAR GRAPHS, AND LINE GRAPHS

Example 1: The pictograph below shows the annual number of house renovations made by one company for the years 1994-1999.

House Renovations

1999	△△△△△△
1998	△△△△△
1997	△△
1996	△△△△
1995	△△△△
1994	△△△

Each △ represents 5 houses

a. How many houses did the company renovate during the year 1996?

b. During which year were the least number of renovations made?

Solution: a.   (number of △) · (5 houses)

             = 4 · 5 houses

             = 20 houses

         20 houses were renovated during 1996.

      b.   The least number of symbols appear next to 1997. This is the year in which the least number of renovations were made.

Example 2:   The following bar graph shows the profits (in dollars) made by ABC Company during the years 1996 to 1999.

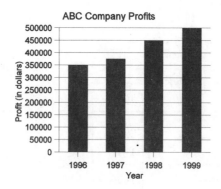

      a.   Approximate the profits for the year 1996.

      b.   Are profits increasing or decreasing?

Solution: a.   Find the year 1996, go to the top of the bar and then go across to the vertical scale to read the dollar amount of profits. The profit for the year 1996 is $350,000.

      b.   As the years increase, the bars are getting taller. Hence the profits are increasing.

Example 3:   The following line graph shows the temperature highs (in degrees Fahrenheit) for one week during the month of September in Manassas.

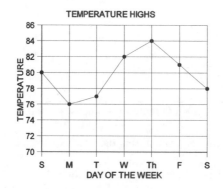

    a.   What was the high temperature reading on Monday?

    b.   What day was the temperature the highest? What was this high temperature?

Solution: a.   Find Monday on the horizontal scale, go straight up until you reach the line graph then go across to the vertical scale and read the temperature. The high temperature reading on Monday was 76° F.

    b.   Locate the highest point on the line graph then go down from this point to the horizontal scale and identify the day. Thursday had the highest temperature. Now go from the point on the line graph across to the vertical scale and identify the temperature. The high temperature for Thursday was 84° F.

8.2 EXERCISES

Using the pictograph from Example 1, do the following exercises.

1. How many houses did the company renovate during the year 1995?

2. How many houses did the company renovate during the year 1998?

3. During which year were the greatest number of renovations made?

4. How many more renovations were made in 1998 than in 1994?

5. Over the years 1994-1999, what was the total number of renovations made?

6. In what year(s) were 20 renovations made?

Using the bar graph from Example 2, do the following exercises.

7. Approximate the profits for the year 1997.

8. During what year were the profits $450,000?

9. How much more profit was made during 1998 than in 1996?

10. Between which two consecutive years was the increase in profit the greatest?

11. Over the years 1996-1999, what was the total profit?

Using the line graph from Example 3, do the following exercises.

12. What was the high temperature reading on Tuesday?

13. What day was the temperature the lowest?

14. Between which two consecutive days was there the greatest increase in high temperature?

15. On what day(s) was the high temperature reading 82° F?

16. How many degrees difference was there between the high temperature reading on Sunday and on Saturday?

8.3 RECTANGULAR COORDINATE SYSTEM

Example 1: Plot each point corresponding to the ordered pairs on the same set of axes.

$$(3, -1), \ (-4, -2), \ (0, 5), \ (-1, 0), \ (2, 3)$$

Solution:

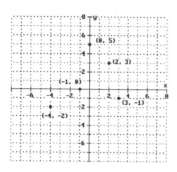

Example 2: Is $(2, -5)$ a solution of the equation $-x + 3y = -17$?

Solution:

$-x + 3y = -17$	Original equation.
$-2 + 3(-5) = -17$	Replace x with 2 and y with -5.
$-2 - 15 = -17$	Multiply.
$-17 = -17$	True.

Since $-17 = -17$ is true, then $(2, -5)$ is a solution of the equation $-x + 3y = -17$.

Example 3: Complete the ordered pair solutions of the equation $2x + 4 = y$.

    a. $(0, \ )$ b. $(\ , 6)$ c. $(-1, \ )$

Solution:

a.

$2x + 4 = y$	Original equation.
$2(0) + 4 = y$	Replace x with 0.
$4 = y$	Solve for y.

The ordered pair solution is $(0, 4)$.

b.

$2x + 4 = y$	Original equation.
$2x + 4 = 6$	Replace y with 6.
$2x + 4 - 4 = 6 - 4$	Solve for x.
$2x = 2$	

$$\frac{2x}{2} = \frac{2}{2}$$

$$x = 1$$

The ordered pair solution is $(1, 6)$.

c. $2x + 4 = y$ Original equation.
    $2(-1) + 4 = y$ Replace x with -1.
     $-2 + 4 = y$ Solve for y.
       $2 = y$

The ordered pair solution is $(-1, 2)$.

8.3 EXERCISES

Plot points corresponding to the ordered pairs on the same set of axes.

1. (6, 2)   2. (-3, -2)   3. (-4, 3)

4. (0, -1)   5. (2, -2)   6. (5, 0)

Find the x- and y-coordinates of each labeled point.

7.

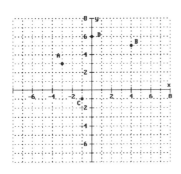

Determine whether each ordered pair is a solution of the given linear equation.

8. $y = 8x$;   (-1, 8)   9. $x = -4y$;   (0, 0)

10. $x + 3y = 5$;   (1, 2)   11. $3x - 5y = 8$;   (1, -1)

Plot the three ordered-pair solutions of the given equation.

12. $-x + y = 4$;   (0, 4),   (-4, 0),   (-2, 2)   13. $-x = 3y$;   (0, 0),   (3, -1),   (-3, 1)

Complete the ordered-pair solutions of the given equations.

14. $x = 10y$;   (10,),   (, 0),   (, 2)   15. $4x - y = 8$;   (2,),   (3,),   (, 8)

16. $x + 6y = 0$;   (, 1),   (, -2),   (0,)

8.4 GRAPHING LINEAR EQUATIONS

Example 1: Graph the equation $-4x = y$ by plotting the following points that satisfy the equation and drawing a line through the points.

$(-1, 4), \quad (0, 0), \quad (1, -4)$

Solution: Plot the points and draw a line through them. The line is the graph of the linear equation $-4x = y$.

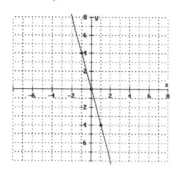

Example 2: Graph $y = 2x + 1$.

Solution:

x	$y = 2x + 1$
-1	$2(-1) + 1 = -1$
0	$2(0) + 1 = 1$
1	$2(1) + 1 = 3$

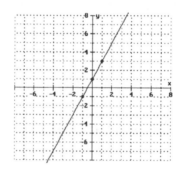

Example 3: Graph $x = 3$.

Solution: The graph is a vertical line that crosses the x-axis at 3.

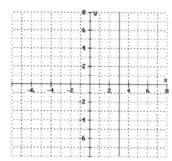

8.4 EXERCISES

Graph each equation.

1. $x + y = -5$
2. $x + 2y = 6$
3. $3x - y = 6$
4. $y - x = -1$

5. $y = 2x - 4$
6. $x = 3y + 1$
7. $y = x + 3$
8. $x = y + 7$

9. $x = 6$
10. $y = -4$
11. $y = 7$
12. $x = -5$

13. $x - 2 = 0$
14. $y + 4 = 0$
15. $x - 4y = 0$
16. $y + 3x = 0$

8.5 MEAN, MEDIAN, AND MODE

Example 1: Find the mean of the following test scores. 72, 86, 76, 92, 85, and 91.

Solution: To find the mean, find the sum of the number items and divide by 6, the number of items.

$$\text{mean} = \frac{72 + 86 + 76 + 92 + 85 + 91}{6} = \frac{502}{6} = 83.7$$

The mean test score is 83.7.

Example 2: Find the median of the following list of numbers. 74, 63, 91, 87, 55, 73, 76

Solution: First, list the scores in numerical order and then find the middle number.

55, 63, 73, <u>74</u>, 76, 87, 91

The median is 74.

Example 3: Find the mode of the following list of numbers.

21, 27, 32, 23, 27, 21, 31, 23, 18, 34, 27

Solution: Find the number that occurs most often. The mode is 27.

8.5 EXERCISES

Find the mean for each of the following lists of numbers.

1. 18, 24, 16, 31, 19, 26, 30
2. 51, 40, 80, 71, 62, 54, 95, 48

3. 9.4, 8.6, 11.2, 17.8, 10.1, 3.4
4. 122, 146, 130, 124, 148, 132, 120, 160, 154, 155

Find the median for each of the following lists of numbers.

5. 0.1, 0.3, 0.7, 0.6, 0.2, 0.9
6. 62, 81, 9, 27, 54, 71, 11, 90, 65

7. 576, 419, 328, 637, 505, 491, 387
8. 80, 60, 90, 75, 105, 45, 87, 91

Find the mode for each of the following lists of numbers.

9. 11, 15, 9, 11, 15, 10, 20, 15, 10, 8, 9, 15   10. 140, 142, 136, 138, 138, 138, 140, 151, 140, 136, 140

11. 5.4, 3.7, 2.1, 4.5, 4.5, 5.4, 7.3, 5.4, 4.5   12. 22, 32, 14, 16, 22, 9, 14, 8

The grades are given for a student for a particular semester. Find the grade point average. If necessary round the grade point average to the nearest hundredth.

13.

Grade	Credit Hours
A	4
C	3
B	3
B	4

14.

Grade	Credit Hours
C	3
C	5
A	3
A	4
B	3

15.

Grade	Credit Hours
B	3
D	3
C	5
D	4
F	3

8.6 COUNTING AND INTRODUCTION TO PROBABILITY

Example 1: Draw a tree diagram for the following experiment. Then use the diagram to find the number of possible outcomes.

Choose a letter a, b, c, d, and then a number 1 or 2.

Solution:

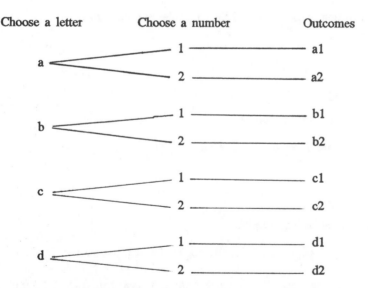

Choose a letter	Choose a number	Outcomes

There are 8 possible outcomes.

Example 2: Using the experiment in Example 1, find the probability of getting a1 or d2.

Solution: probability of a1 or d2 $= \dfrac{2}{8}$ → number of ways event can occur

    → number of possible outcomes

$= \dfrac{1}{4}$ simplest form

8.6 EXERCISES

Draw a tree diagram for each experiment. Then use the diagram to find the number of possible outcomes.

1. Choose a number 1, 2, 3, or 4 and then toss a coin.

2. Toss a coin and then choose a number 1, 2, 3, or 4.

3. Roll a die and then choose a vowel a, e, i, o, u.

4. Toss a coin and then choose a number 5, 6, or 7.

If a single choice is made from a bag with 8 purple marbles, 5 orange marbles, 3 pink marbles, and 4 white marbles, find the probability of each event.

5. A white marble is chosen

6. A pink marble is chosen.

7. A purple marble is chosen.

8. An orange marble is chosen.

CHAPTER 8 PRACTICE TEST

The circle graph below shows the age groups of students enrolled in a mathematics course during a particular term. There was a total of 600 students enrolled in a mathematics course during that term.

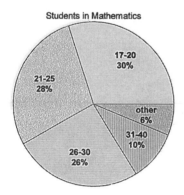

Students in Mathematics

1. Which age group contains the most students?

2. Find the number of students that were in the 26-30 age range.

3. Which age group contained 10% of the students?

The pictograph below shows the money taken in each week from a popcorn fundraiser.

Weekly Popcorn Sales

1	○ ○ ○ ○
Weeks 2	○ ○ ○ ○ ○ ○
3	○ ○ ○
4	○ ○ ○ ◖
5	○ ○ ◖
6	○

Each ○ represents $50

4. How much money was taken in during the third week?

5. During which week was the most money taken in? How much money was taken in during that week?

6. During which week was the least amount of money taken in? How much money was taken in during that week?

7. What was the total money taken in for the fundraiser?

Find the coordinates of each point.

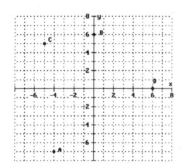

8. A   9. B   10. C   11. D

Complete and graph the ordered-pair solutions of the given equation.

12. $x + 2y = 1$; (, 0), (3,), (−3,)   13. $y = 5x - 2$; (0,), (1,) (, −7)

Graph each linear equation.

14. $x - y = 2$ 15. $x + 4 = 0$ 16. $y - 1 = 0$ 17. $5x + 6y = 30$

Find the mean, median, and mode of each list of numbers.

18. 61, 60, 72, 63, 59, 74   19. 101, 94, 113, 89, 126, 94, 112

90

Find the grade point average. If necessary, round to the nearest hundredth.

20.

Grade	Credit Hours
D	3
B	5
B	3
A	4

21. If a die is rolled one time, find the probability of rolling a 5 or 6.

22. If a coin is tossed twice, find the probability of tossing a tail and then a tail.

CHAPTER 8 HELPFUL HINTS AND INSIGHTS

- Remember that **each point** in the rectangular coordinate system corresponds to exactly **one ordered pair** and that **each ordered pair** corresponds to exactly **one point.**

- Since the first number or x-coordinate of an ordered pair is associated with the x-axis it tells how many units to move left or right. Similarly, the second number or y-coordinate tells how many units to move up or down.

- If a is a number then, the graph of $y = a$ is a horizontal line that crosses the y-axis at a.

- If a is a number, then the graph of $x = a$ is a vertical line that crosses the x-axis at a.

9.1 LINES AND ANGLES

Example 1: Classify each angle as acute, right, obtuse, or straight.

a. b.

c. d.

Solution: a. ∠A is an obtuse angle. It measures between 90° and 180°.

b. ∠B is a right angle.

c. ∠C is a straight angle.

d. ∠D is an acute angle. It measures between 0° and 90°.

Example 2: Find the complement of a 13° angle.

Solution: The complement of a 13° angle is an angle that measures 90° − 13° = 77°.

Example 3: Find the measure of ∠*a*.

Solution: Since the sum of the measures of the three angles is 180°, measure of
∠*a* = 180° − 120° − 25° = 35°

9.1 EXERCISES

Identify each figure as a line, a ray, a line segment, or an angle.

1. 2.

3. 4.

Find the measure of each angle in the figure.

5. ∠ABD   6. ∠EBC   7. ∠CBD

Classify each angle as acute, right, obtuse, or straight.

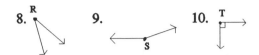

8. R   9.   10. T

11. Find the complement of a 64° angle.

12. Find the supplement of an 88° angle.

13. Find the measures of angles x, y, and z.

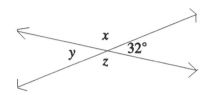

Find the measure of ∠ x in each figure.

14.   15.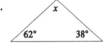

9.2 LINEAR MEASUREMENT

Example 1: Convert 11 feet to inches.

Solution: $11 \text{ ft} = 11 \text{ ft} \cdot \dfrac{12 \text{ in.}}{1 \text{ ft}}$

           $= 11 \cdot 12 \text{ in.}$

           $= 132 \text{ in.}$

Example 2: Convert 6 yards 3 feet to feet.

Solution: $6 \text{ yards} = 6 \text{ yds} \cdot \dfrac{3 \text{ ft}}{1 \text{ yd}}$

           $= 6 \cdot 3 \text{ ft}$
           $= 18 \text{ ft}$

   $6 \text{ yards } 3 \text{ ft} = 18 \text{ ft} + 3 \text{ ft} = 21 \text{ ft}$

Example 3: Add 5 ft 4 in. and 7 ft 9 in.

Solution:
```
      5 ft   4 in.
  +   7 ft   9 in.
     12 ft  13 in.
  = 12 ft 1 ft 1 in.
  = 13 ft 1 in.
```

Example 4: Convert 36,000 cm to kilometers.

Solution:

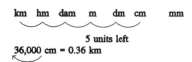

9.2 EXERCISES

Convert each measurement as indicated.

1. 75 in. to feet

2. 17 yd to feet

3. 42,240 ft to miles

4. 2.6 mi to feet

5. 82 ft to yards

6. 97 ft to inches

7. 38 ft = —— yd —— ft

8. 83 in. = —— ft —— in.

9. 20,000 ft = —— mi —— ft

10. 6 ft 4 in. = —— in.

11. 12 yd 2 ft = —— ft

12. 65 m to centimeters

13. 5800 mm to meters

14. 19.6 mm to decimeters

15. 0.9 m to millimeters

Perform the indicated operations.

16. 11 ft 6 in. + 8 ft 9 in

17. 25 ft 3 in. – 10 ft 8 in.

18. 32 yd 2 ft × 5

19. 20 cm – 18 mm

20. 12.8 m ÷ 4

9.3 PERIMETER AND PROBLEM SOLVING

Example 1: Find the perimeter of the rectangle below.

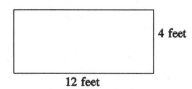

Solution: Perimeter = 12 feet + 12 feet + 4 feet + 4 feet
                     = 32 feet

The perimeter of the rectangle is 32 feet.

Example 2: Find the perimeter of a square with side length 6 inches.

Solution: $P = 4s$
                   $= 4(6 \text{ inches})$ Let $s = 6$ inches.
                   $= 24 \text{ inches}$
The perimeter of the square is 24 inches.

Example 3: Find the perimeter of a triangle if the sides are 6 meters, 9 meters and 14 meters.

Solution: $P = a + b + c$
                   $= 6 \text{ meters} + 9 \text{ meters} + 14 \text{ meters}$
                   $= 29 \text{ meters}$
The perimeter of the triangle is 29 meters.

Example 4: Find how much fencing is needed to enclose a rectangular vegetable garden 90 feet by 22 feet.

Solution: $P = 2l + 2w$
                   $= 2(90 \text{ feet}) + 2(22 \text{ feet})$
                   $= 180 \text{ feet} + 44 \text{ feet}$
                   $= 224 \text{ feet}$
It will take 224 feet of fencing to enclose the garden.

9.3 EXERCISES

Find the perimeter of each figure.

1.

Square   11 meters

2.

10 inches
3 inches   9 inches

3.

Rectangle   5 feet
16 feet

4.

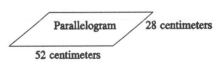

Parallelogram   28 centimeters
52 centimeters

5.

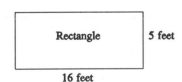

7 feet   2 feet
4 feet
9 feet
13 feet

6.

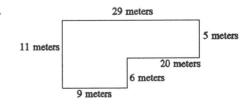

29 meters
11 meters   5 meters
20 meters
6 meters
9 meters

Solve.

7. A polygon has sides of length 14 feet, 18 feet, 9 feet, 7 feet and 3 feet. Find its perimeter.

8. A triangle has sides of 20 centimeters, 30 centimeters and 50 centimeters. Find its perimeter.

9. A square has a side of length 15 miles. Find its perimeter.

10. A rectangle has dimensions 31 yards by 42 yards. Find its perimeter.

11. A metal strip is being installed around a cabinet that is 6 feet long and 2 feet wide. Find how much stripping is needed.

12. Find how much fencing is needed to enclose a rectangular plot 92 feet by 28 feet.

13. The perimeter of a rectangular field is 600 meters. If the field is four times longer than it is wide, find the length of the field.

14. Two sides of a triangle are the same length. The third side is 33 inches. If the perimeter of the triangle is 121 inches, find the length of each equal side.

15. The width of a rectangle is 12 meters less than twice its length. If the perimeter is 30 meters, find its width.

16. Find the perimeter of a square floor tile with a side of 8 inches.

9.4 AREA AND VOLUME

Example 1: Find the area of the triangle.

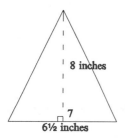

Solution: Area $= \dfrac{1}{2} \cdot$ base $\cdot$ height

$$= \frac{1}{2}\left(6\frac{1}{2} \text{ in.}\right) \cdot (8 \text{ in.})$$

$$= \frac{1}{2} \cdot \frac{13}{2} \cdot \frac{8}{1} \text{ sq. in.}$$

$$= \frac{1 \cdot 13 \cdot 2 \cdot 2 \cdot 2}{2 \cdot 2 \cdot 1} \text{ sq. in.}$$

$$= 26 \text{ sq. in.}$$

Example 2: Find the area of a circle with a radius of 8 centimeters. Find the exact area and an approximation. Use 3.14 as an approximation for π.

Solution: Area $= \pi r^2$
            $= \pi (8 \text{ cm})^2$ Let $r = 8$ cm
            $= 64\pi$ sq. cm

To approximate this area, we substitute 3.14 for π.

64π sq. cm $\approx 64(3.14)$ sq. cm
            $= 200.96$ sq. cm

The **exact** area of the circle is 64π sq. cm, which is **approximately** 200.96 sq. cm.

Example 3: Find the volume of a rectangular box that is 18 centimeters long, 9 centimeters wide, and 4 centimeters high.

Solution: Volume $=$ length $\cdot$ width $\cdot$ height
            $= (18 \text{ cm}) \cdot (9 \text{ cm}) \cdot (4 \text{ cm})$
            $= 648$ cubic cm

9.4 EXERCISES

Find the area of the geometric figure. If the figure is a circle, give an exact area and then use the given approximation for π to approximate the area.

1.

2.
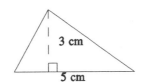

3.
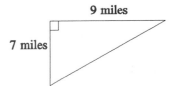

4.
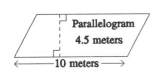

5.
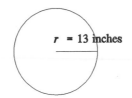

Use 22/7 for π.

6.

Use 3.14 for π.

7.
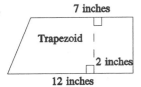

8.
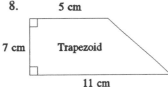

9.
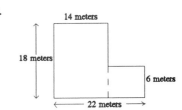

Find the volume of the solid. Use 3.14 as an approximation for π.

10. rectangular box with length = 2 inches, width = 8 inches, and height = 3 inches

11. sphere with radius 28 miles

12. circular cylinder with base radius 2.2 feet and height 16 feet

13. cone with base radius 5 meters and height 19 meters

14. square-based pyramid with side 4.7 inches and height 18.6 inches

Solve.

15. A $14\frac{1}{2}$-foot by 20-foot concrete wall is to be built using concrete blocks. Find the area of the wall.

16. A picture frame measures 14 inches by $8\frac{1}{2}$ inches. Find how many square inches of glass the frame requires.

17. Find the volume of a cube with edges $2\frac{3}{4}$ inches.

18. Find the volume of a rectangular block of ice 3 feet by $1\frac{1}{2}$ feet by $\frac{1}{2}$ foot.

9.5 WEIGHT AND MASS

Example 1: Convert 928 ounces to pounds.

Solution: $928 \text{ oz} = 928 \text{ oz} \cdot \dfrac{1 \text{ lb}}{16 \text{ oz}} = \dfrac{928}{16} \text{ lb}$

$$= 58 \text{ lb}$$

Example 2: Multiply 4 lb 7 oz by 5.

Solution:
```
    4 lb   7 oz
  ×          5
  20 lb  35 oz

         2 lb 30z
    16)35
       32
        3
```

Thus 20 lb 35 oz = 20 lb + 2 lb 3 oz = 22 lb 3 oz.

Solution: Convert both numbers to milligrams or grams before subtracting.

58 mg = 0.058 g   or   7.9 g = 7900 mg

```
    7.900 g            7900 mg
  − 0.058 g          −   58 mg
    7.842 g            7842 mg
```

The difference is 7.842 g   or   7842 mg.

9.5 EXERCISES

Convert the following as indicated.

1. 5 pounds to ounces

2. 6 tons to pounds

3. 7200 pounds to tons

4. 22.25 pounds to ounces

5. 82 ounces to the nearest tenth of a pound

6. 6000 g to kilograms

7. 12 g to milligrams

8. 7.2 g to kilograms

9. 6.92 kg to grams

10. 8025 cg to hectograms

Perform the indicated operations.

11. 19 lb 4 oz + 31 lb 15 oz

12. 1 ton 1905 lb + 5 ton 168 lb

13. 14 lb 3 oz − 4 lb 8 oz

14. 3 lb 7 oz × 9

15. 6 tons 300 lb ÷ 5

16. 89.7 g + 10.6 g

17. 6 kg − 3560 g

18. 2.9 kg × 8.6

Solve.

19. One can of kidney beans weighs 15 oz. How much will 9 cans weigh?

20. Mike normally weighs 105 kg, but he lost 20,000 grams after dieting. Find Mike's new weight.

9.6 CAPACITY

Example 1: Convert 13 quarts to gallons.

Solution: $13 \text{ qt} = 13 \text{ qt} \cdot \dfrac{1 \text{ gal}}{4 \text{ qt}}$

$= \dfrac{13}{4} \text{ gal}$

$= 3\dfrac{1}{4} \text{ gal}$

Example 2: Add 2 qt + 5 gal 3 qt

Solution:

$$
\begin{array}{r}
2 \text{ qt} \\
+\ 5 \text{ gal} \quad 3 \text{ qt} \\
\hline
5 \text{ gal} \quad 5 \text{ qt} \\
=\ 5 \text{ gal} \quad 1 \text{ gal} \ 1 \text{ qt} \\
=\ 6 \text{ gal} \quad 1 \text{ qt}
\end{array}
$$

Example 3: Multiply 12.4 L × 5.

Solution:

$$
\begin{array}{r}
12.4 \text{ L} \\
\times\quad 5 \\
\hline
62.0 \text{ L}
\end{array}
$$

9.6 EXERCISES

Convert each measurement as indicated.

1. 28 quarts to gallons

2. 11 quarts to pints

3. 18 cups to pints

4. 40 cups to gallons

5. $5\frac{3}{4}$ quarts to cups

6. $5\frac{7}{8}$ gallons to pints

7. 9 L to milliliters

8. 3200 ml to liters

9. 140 L to kiloliters

10. 2.5 kl to liters

Perform the indicated operations.

11. 8 gal 2 qt + 6 gal 1 qt

12. 7 c 4 fl oz + 3 c 6 fl oz

13. 4 gal 1 pt × 3

14. 3 pt − 2 pt 1 c

15. 20.2 L + 14.6 L

16. 7920 ml − 0.2 L

17. 6.4 L ÷ 0.8

18. 125 ml × 7

Solve.

19. A recipe for crepes calls for $1\frac{1}{2}$ cups of milk. How many fluid ounces is this?

20. Terry wants to divide a 1-L bottle of juice equally between her 4 children. How much will each child get?

9.7 TEMPERATURE

Example 1: Convert 35°C to degrees Fahrenheit.

Solution: $F = \dfrac{9C}{5} + 32$ Use the conversion formula.

$= \dfrac{9 \cdot 35}{5} + 32$ Replace C with 35.

$= 63 + 32$ Simplify.
$= 95$ Add.

Thus, 35°C is equivalent to 95°F.

Example 2: Convert 62° F to degrees Celsius. If necessary, round to the nearest tenth of a degree.

Solution: $C = \dfrac{5(F - 32)}{9}$ Use the conversion formula.

$= \dfrac{5(62 - 32)}{9}$ Replace F with 62.

$= \dfrac{5(30)}{9}$

$= 16.\overline{6}$

Hence, 62°F is equivalent to 16.7°C.

9.7 EXERCISES

Convert the following as indicated. When necessary, round to the nearest tenth of a degree.

1. 85°C to degrees Fahrenheit

2. 30°C to degrees Fahrenheit

3. 113°F to degrees Celsius

4. 77°F to degrees Celsius

5. 105°C to degrees Fahrenheit

6. 58°F to degrees Celsius

7. 49°F to degrees Celcius

8. 80°C to degrees Fahrenheit

9. 22.4°C to degrees Fahrenheit

10. 121.7°F to degrees Celsius

Solve.

11. Carolyn noticed that her outdoor thermometer was registering a temperature of 95° F. Convert this measurement to degrees Celsius.

12. Chi's office is normally kept at a temperature of 70° F. Convert this measurement to degrees Celsius.

13. A weather forecaster in Helsinki predicts a high temperature of 20° C. Find the measurement in degrees Fahrenheit.

14. The sign at the bank indicates the current temperature to be 12° C. Convert this measurement to degrees Fahrenheit.

15. Moojan is running a fever of 102.3° F. Find her temperature as it would be shown on a Celsius thermometer.

CHAPTER 9 PRACTICE TEST

1. Find the complement and supplement of a 57° angle.

2. Find the measures of angles *x, y,* and *z.*

3. Find the measure of ∠*x.*

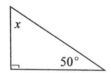

4. Find the perimeter of a rectangle with width 19 feet and length 30 feet.

5. A poster has a perimeter of 120 inches. The length of the poster is 12 inches more than its width. Find the dimensions of the poster.

6. Find the area of a rectangle with length 11 inches and width 4.5 inches.

7. Find the volume of a cube with edges 8¼ meters.

8. How much potting soil is needed to fill a flower box 3 feet by 1½ feet by ½ foot?

Convert as indicated.

9. 140 in. to feet and inches

10. 26 qt to gallons

11. 46 oz to pounds

12. 4.6 tons to pounds

13. 42 pt to gallons

14. 62 mg to grams

15. 9.8 kg to grams

16. 7.2 cm to millimeters

17. 9.1 dg to grams

18. 0.075 L to milliliters

Perform the indicated operations.

19. 5 qt 1 pt + 7 qt 1 pt

20. 10 lb 5 oz - 3 lb 7 oz

21. 3 ft 4 in. × 5

22. 8 gal 1 qt ÷ 3

23. 11 cm - 16 mm

24. 2.4 km + 329 m

Convert. Round to the nearest tenth of a degree, if necessary.

25. Convert 74°F to degrees Celsius.

26. Convert 10.8°C to degrees Fahrenheit.

CHAPTER 9 HELPFUL HINTS AND INSIGHTS

- A millimeter is about the thickness of a large paper clip wire.
  A centimeter is about the width of a large paper clip.
  A meter is slightly longer than a yard.
  A kilometer is about two-thirds of a mile.

- A kilogram is slightly over 2 pounds.
  A large paper clip weighs approximately 1 gram.

- A liter of liquid is slightly more than one quart.

- Area is always measured in square units.

- When finding the area of figures, be sure all measures are changed to the same unit before calculations are made.

- Volume is always measured in cubic units.

10.1 ADDING AND SUBTRACTING POLYNOMIALS

Example 1: Add: $(5x + 3) + (-8x + 7)$

Solution: $(5x + 3) + (-8x + 7) = (5x - 8x) + (3 + 7)$ Group like terms.
$$= -3x + 10$$ Combine like terms.

Example 2: Subtract $(4y^2 - 3y + 12)$ from $(9y^2 - 11)$.

Solution:

$$\begin{array}{r} 9y^2 \qquad\;\; - 11 \\ - (4y^2 - 3y + 12) \end{array}$$

becomes
$$\begin{array}{r} 9y^2 \qquad\;\; - 11 \\ - 4y^2 + 3y - 12 \\ \hline 5y^2 + 3y - 23 \end{array}$$

Example 3: Find the value of the polynomial $-5z^2 - z + 18$ for $z = -2$.

Solution: $-5z^2 - z + 18 = -5(-2)^2 - (-2) + 18$ Let $z = -2$.
$$= -5(4) + 2 + 18$$
$$= -20 + 2 + 18$$
$$= 0$$

10.1 EXERCISES

Perform the following operations.

1. $(3x - 2) + (-9x + 31)$

2. $(18y + 6) + (10y - 22)$

3. $(2t + 9) + (7t^2 - 8t + 5)$

4. $(30x + 5) - (25x - 2)$

5. $(-18x^2 + 4x - 1) - (-2x + 17)$

6. $(10y^3 + 15y^2 - y - 2) - (8y^3 + 6y - 12)$

7. $(3.7b^3 + 20) + (-5.3b^2 - 6.1b + 7)$

8. Subtract $(16t - 7)$ from $(-9t - 7)$.

9. Subtract $(x^2 - 6x + 5)$ from $(21x^2 + 5x - 5)$.

10. Subtract $\left(4x^2 - \dfrac{3}{8}x\right)$ from $\left(-3x^2 + \dfrac{5}{8}x\right)$.

Find the value of each polynomial for $x = 3$.

11. $-9x - 2$

12. $4x + 13$

13. $x^2 - 2x + 5$

14. $-x^2 - x - 1$

15. $\dfrac{4x^2}{6} + 10$

16. $\dfrac{x^3}{9} - x - 11$

Find the value of each polynomial for $x = -4$.

17. $4x - 1$

18. x^2

19. x^3

20. $x^3 - x^2 + x + 1$

10.2 MULTIPLICATION PROPERTIES OF EXPONENTS

Example 1: Multiply: $x^5 \cdot x^4$

Solution: $x^5 \cdot x^4 = x^{5+4}$
$$= x^9$$

Example 2: Multiply: $(-4a^3b^7)(-2ab^8)$

Solution: $(-4a^3b^7)(-2ab^8) = (-4)(-2)(a^3 \cdot a^1)(b^7 \cdot b^8)$
$$= 8a^{3+1}b^{7+8}$$
$$= 8a^4b^{15}$$

Example 3: Simplify: $(y^{12})^4$

Solution: $(y^{12})^4 = y^{12 \cdot 4}$
$$= y^{48}$$

Example 4: Simplify: $(-2x^2y^3)^5 \cdot (3x^3y^4)^6$

Solution: $(-2x^2y^3)^5 \cdot (3x^3y^4)^6 = (-2)^5(x^2)^5(y^3)^5 \cdot (3)^6(x^3)^6(y^4)^6$
$$= -32x^{10}y^{15} \cdot 729x^{18}y^{24}$$
$$= (-32 \cdot 729)(x^{10} \cdot x^{18})(y^{15} \cdot y^{24})$$
$$= -23328x^{28}y^{39}$$

10.2 EXERCISES

Multiply and simplify.

1. $x^4 \cdot x^9$

2. $b^{12} \cdot b$

3. $2y^3 \cdot 6y^4$

4. $-8x \cdot 11x$

5. $(-10a^2b)(-2a^6b^5)$

6. $(-x^3y^5z)(-3xy^2z^4)$

7. $2x \cdot 6x \cdot x^2$

8. $a \cdot 9a^{10} \cdot 10a^9$

9. $(x^{10})^9$

10. $(y^{15})^3$

11. $(2m)^4$

12. $(y^6)^3 \cdot (y^2)^4$

13. $(x^3y^8)^7$

14. $(8m^5n^{12})^3$

15. $(-4z)(2z^9)^4$

16. $(2xy^2)^5(3x^6y^4)^3$

17. $(14y^{12}z^8)^2$

18. $(4a^3b^{12})^3(2a^7b)^4$

Find the area of each figure.

19.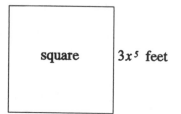

$3x^5$ feet

20. $2y^4$ meters

rectangle $2y$ meters

10.3 MULTIPLYING POLYNOMIALS

Example 1: Multiply: $5x(2x^2 + 6)$

Solution: $5x(2x^2 + 6) = 5x \cdot 2x^2 + 5x \cdot 6$ Apply the distributive property.

$$= 10x^3 + 30x$$

Example 2: Multiply: $(x + 4)(x + 7)$

Solution: $(x + 4)(x + 7) = x(x + 7) + 4(x + 7)$ Apply the distributive property.
$= x \cdot x + x \cdot 7 + 4 \cdot x + 4 \cdot 7$ Apply the distributive property.
$= x^2 + 7x + 4x + 28$ Multiply.
$= x^2 + 11x + 28$ Combine like terms.

Example 3: Multiply: $(5x - 2)^2$

Solution: $(5x - 2)^2 = (5x - 2)(5x - 2)$ Apply the definition of an exponent.
$= 5x(5x - 2) - 2(5x - 2)$ Apply the distributive property.
$= 5x \cdot 5x + 5x \cdot (-2) - 2 \cdot 5x - 2 \cdot (-2)$ Apply the distributive property.
$= 25x^2 - 10x - 10x + 4$ Multiply
$= 25x^2 - 20x + 4$ Combine like terms.

Example 4: Multiply: $(2y + 7)(y^2 + 4y - 3)$

Solution: $(2y + 7)(y^2 + 4y - 3)$

$= 2y(y^2 + 4y - 3) + 7(y^2 + 4y - 3)$ Apply the distributive property.

$= 2y \cdot y^2 + 2y \cdot 4y + 2y(-3) + 7 \cdot y^2 + 7(4y) + 7(-3)$ Apply the distributive property.

$= 2y^3 + 8y^2 - 6y + 7y^2 + 28y - 21$ Multiply.

$= 2y^3 + 15y^2 + 22y - 21$ Combine like terms.

10.3 EXERCISES

Multiply.

1. $7x(8x^2 + 5)$ 2. $3y(9y^4 - y)$ 3. $-6m(3m^2 - m - 2)$

4. $8t^3(-2t^2 + t + 6)$

5. $(x - 3)(x - 9)$

6. $(y + 10)(y + 1)$

7. $(m + 8)(m - 8)$

8. $(3x + 5)(x - 11)$

9. $(2t + 5)(2t + 7)$

10. $(4y - 5)^2$

11. $(3x + 8)^2$

12. $\left(z + \dfrac{1}{4}\right)\left(z - \dfrac{3}{4}\right)$

13. $(4a + 9)(4a - 9)$

14. $(x + 7)^2$

15. $(a - 2)(6a^2 + 5a + 13)$

16. $(x + 5)(x^2 - 5x + 25)$

17. $(y^2 + 3)(2y^2 - y + 8)$

18. $(x^2 + 2x + 3)(x^3 - x^2 + 4)$

Find the area of each figure.

19.

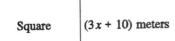

Square $(3x + 10)$ meters

20.

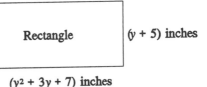

Rectangle $(y + 5)$ inches

$(y^2 + 3y + 7)$ inches

10.4 INTRODUCTION TO FACTORING POLYNOMIALS

Example 1: Find the GCF of $8x^4$, $12x^2$, and $36x^3$.

Solution: The GCF of 8, 12, and 36 is 4.

The GCF of x^4, x^2, and x^3 is x^2.

Thus the GCF of $8x^4$, $12x^2$, and $36x^3$ is $4x^2$.

Example 2: Factor: $5x^2 + 35x - 25$

Solution: The GCF of the terms is 5.

$$5x^2 + 35x - 25 = 5 \cdot x^2 + 5 \cdot 7x - 5 \cdot 5$$
$$= 5(x^2 + 7x - 5)$$

Example 3: Factor: $-10y^2 + 6y + 8$

Solution: The GCF of the terms is 2.

$$-10y^2 + 6y + 8 = 2 \cdot (-5y^2) + 2 \cdot 3y + 2 \cdot 4$$
$$= 2(-5y^2 + 3y + 4)$$

10.4 EXERCISES

Find the GCF of each list of terms.

1. x^9, x^4, x^7

2. $2y^2, 8y^5, 6y$

3. a^2b^3, a^3b^2, ab^2

4. $3x^3y^4z^2, 15x^2y^3z^3, 6x^3y^4z^2$

5. $14yz, 21y^2z, 35y^3z^2$

6. $18a^4b^5, 27a^6b^7, 54a^7b^5$

Factor.

7. $4x^3 + 8x^2$

8. $6y^4 - 3y$

9. $20z^4 + 10z^2$

10. $12a^3 - 4a^5$

11. $b^{12} - 2b^8$

12. $m^{10} + 5m^4$

13. $13x^6 - 26x^4 + 39x^2$

14. $8y^5 - 4y^3 - 12y^2$

15. $3a^4 - 6b^2 + 9$

16. $11m^4 + 22m^3 - 33m$

CHAPTER 10 PRACTICE TEST

Add or subtract as indicated.

1. $(15x - 7) + (2x + 8)$

2. $(15x - 7) - (2x + 8)$

3. $(4.9y^2 + 10) + (3.2y^2 - 5y - 17)$

4. Subtract $(6b^2 - 5)$ from $(4b^2 - 4b - 1)$.

5. Find the value of $x^2 + 3x - 5$ when $x = 6$.

Multiply and simplify.

6. $y^6 \cdot y^8$

7. $(x^5)^9$

8. $(3x^6)^3$

9. $(-15y^2)(-4y^8)$

10. $(b^{12})^5(b^2)^3$

11. $(4x^3y)^4(3yx^2)^3$

12. $6x(9x^3 - 2.6)$

13. $-5a(a^4 - 10a^3 + 1)$

14. $(x - 6)(x - 7)$

15. $(4x + 9)^2$

16. $(y - 7)(y^2 + 7y + 49)$

17. Find the area and the perimeter of a rectangle with length $(3x + 4)$ feet and width $5x$ feet.

Find the greatest common factor of the list of terms.

18. 75 and 90

19. $4x^6, 18x^3, 24x^9$

Factor.

20. $7x^2 - 35$ 21. $y^{10} - 4y^6$ 22. $3x^2 - 6x + 30$

23. A rock is dropped from the top of a 150-foot cliff. The height of the rock at time t seconds is given by the polynomial $-16t^2 + 150$. Find the height of the rock when $t = 2$ seconds.

CHAPTER 10 HELPFUL HINTS AND INSIGHTS

- To find the opposite of a polynomial, change the signs of all the terms of the polynomial.

- Make sure you understand when to use the product property and when to use the power of a power property.

  $x^5 \cdot x^7 = x^{5+7} = x^{12}$ product property

  $(x^5)^7 = x^{5 \cdot 7} = x^{35}$ power of a power property

PRACTICE FINAL EXAMINATION A

Evaluate.

1. $62 + 57$

2. $691 - 328$

3. $127(13)$

4. $2835 \div 81$

5. $(3^2 - 4) \cdot 8$

6. $21 + 12 \div 6 \cdot 4 - 7$

7. Round 2,137,546 to the nearest thousand.

8. Evaluate $4(x^3 - 6)$ if $x = 2$.

9. Twenty-six cans of varnish cost $312. How much was each can?

Simplify each expression.

10. $18 - 45$

11. $-22 + 17$

12. $(-3) \cdot (-30)$

13. $\dfrac{-57}{3}$

14. $(-8)^2 - 24 \div (-6)$

15. $\dfrac{|39 - 57|}{6}$

16. $\dfrac{8(-11) + 12}{-1(9 - 28)}$

17. $\dfrac{25}{-5} - \dfrac{4^3}{8}$

18. Evaluate $3x - y$ for $x = -7$ and $y = -6$.

19. Evaluate $|x - y| + |x|$ for $x = -8$ and $y = 10$.

20. Simplify $7x - 17 - 12x + 30$ by combining like terms.

21. Multiply $-10(2x + 9)$

22. Find the perimeter of a square with a side length of 17 feet.

Solve the following equations.

23. $8x + x = 81$

24. $22 = 3x - 14x$

25. $19x + 12 - 18x - 30 = 20$

26. $5(x + 4) = 0$

27. $10 + 6(2y - 1) = 28$

28. $4(3x - 5) = 2(7x + 8)$

Solve.

29. The sum of three times a number and seven times the same number is ninety. Find the number.

30. Find the length of the side of a square porch with a 72-foot perimeter.

Perform the indicated operations and write the answers in lowest terms.

31. $\dfrac{4}{5} \cdot \dfrac{35}{8}$

32. $\dfrac{5x}{7} - \dfrac{3x}{7}$

33. $\dfrac{xy^3}{z^2} \cdot \dfrac{z^3}{x^4y}$

34. $\dfrac{6y}{13} + \dfrac{5}{26}$

35. $4\dfrac{1}{5} \div \dfrac{7}{25}$

36. $4\dfrac{1}{3} + 3\dfrac{5}{6} + 8\dfrac{2}{9}$

37. $18 \div 3\dfrac{4}{9}$

38. $20y^2 \div \dfrac{y}{5}$

39. $-\dfrac{8}{7} \div \dfrac{64}{21}$

40. $\left(\dfrac{15}{8} \cdot \dfrac{32}{3} \right) \div 6$

41. Simplify $\dfrac{6 + \dfrac{3}{5}}{4 - \dfrac{7}{10}}$.

Solve.

42. $\dfrac{x}{3} + x = -\dfrac{16}{9}$

43. $\dfrac{5}{9} + \dfrac{x}{6} = \dfrac{2}{3} - \dfrac{x}{12}$

44. Evaluate xy for $x = -\dfrac{2}{5}$ and $y = 2\dfrac{3}{10}$.

Perform the indicated operation. Round the result to the nearest thousandth if necessary.

45. $11.654 + 3.71 + 9.827$

46. $-51.6 - 25.82$

47. $(14.6)(3.25)$

48. $(-1.2)^2 + 3.7$

49. $\dfrac{0.11 - 3.75}{0.2}$

50. Round 126.6457 to the nearest hundredth.

51. Write 0.72 as a fraction in simplest form.

52. Write $\dfrac{9}{200}$ as a decimal.

53. Simplify $-2\sqrt{36} + \sqrt{25}$.

54. Find the volume of a cube with edges $2\dfrac{5}{6}$ centimeters.

55. Find the area of a rectangle with length 9 feet and width 2.7 feet.

Complete and graph the ordered-pair solutions of the given equation.

56. $-2x + 3y = 12$; $(0, \quad)$, $(\quad , 0)$, $(\quad , 2)$

57. $y = 5x + 1$; $(0, \quad)$, $(1, \quad)$, $(\quad , -4)$

Graph each linear equation.

58. $x + y = 3$ 59. $x - 5 = 0$ 60. $y + 4 = 0$ 61. $4x - 5y = 20$

62. Find the mean of the following list of numbers: 58, 42, 67, 71, 49, 52, 60

63. Find the median of the following list of numbers: 75, 16, 83, 92, 47, 16, 21, 28, 37

64. Find the mode of the following list of numbers: 31, 16, 90, 81, 16, 80, 17, 16, 31, 30, 45

65. Find the grade point average. If necessary, round to the nearest hundredth.

Grade	Credit Hours
A	3
A	4
B	5
C	3

66. Write the ratio as a fraction in simplest form: 365 bushes to 657 bushes.

67. Write 972 miles in 18 hours as a unit rate.

Solve the proportion for the variable.

68. $\dfrac{15}{x} = \dfrac{105}{98}$

69. $\dfrac{\frac{9}{5}}{\frac{3}{4}} = \dfrac{x}{\frac{25}{18}}$

Convert as indicated.

70. 546 in. to feet and inches

71. 7½ gal to quarts

72. 62 g to milligrams

73. 5700 ml to liters

74. 95°F to degrees Celsius.

Perform the indicated operations.

75. 12 lb 6 oz + 10 lb 14 oz

76. 2.7 km + 231 m

77. 9 cm – 15 mm

78. Write 0.0012 as a percent.

79. Write 32.8% as a decimal.

80. Write $\dfrac{1}{20}$ as a percent.

81. What is 68% of 900?

82. 182 is 35% of what number?

83. 40.32 is what percent of 96?

84. A birdfeed mixture is 22% sunflower seeds. How many pounds of sunflower seeds are contained in 210 pounds of this mixture?

85. If the local sales tax rate is 6.5%, find the total amount charged for a VCR priced at $375.

86. $9600 is compounded quarterly at 6%. Find the total amount in the account after 5 years.

87. Dolores is paid a commission rate of 2.6% of her sales. Find her commission for a week in which her sales were $18,700.

Add or subtract as indicated.

88. $(9y + 15) + (-6y + 8)$

89. $(12x - 5) - (x - 1)$

90. Find the value of $x^2 + x - 7$ when $x = 5$.

Multiply and simplify.

91. $y^{15} \cdot y^{16}$

92. $(x^8)^9$

93. $(a^3)^5(a^{10})^6$

94. $-9y(y^2 - 4y + 3)$

95. $(x + 11)(x - 9)$

96. $(4x + 13)^2$

97. $(b - 8)(b^2 + 8b + 64)$

98. Find the greatest common factor of $12x^7$, $16x^2$, $28x^5$.

Factor.

99. $8x^3 - 64x^5$

100. $10y^2 - 20y$

PRACTICE FINAL EXAMINATION B

Evaluate.

1. $76 + 89$

2. $732 - 468$

3. $212(18)$

4. $5915 \div 65$

5. $(2^3 - 3) \cdot 7$

6. $13 + 18 \div 9 \cdot 5 - 10$

7. Round 561,382 to the nearest hundred.

8. Evaluate $9(x^2 - 12)$ if $x = 7$.

9. Fourteen cases of soda cost $76.86. What is the cost of each case?

Simplify each expression.

10. $21 - 58$

11. $-37 + 10$

12. $(-10) \cdot 8$

13. $\dfrac{-91}{-7}$

14. $(-9)^2 - 35 \div (-5)$

15. $\dfrac{|9 - 60|}{-17}$

16. $\dfrac{4(-12) + 8}{-1(-2 - 6)}$

17. $\dfrac{30}{-6} - \dfrac{3^3}{9}$

18. Evaluate $5x - y$ for $x = -9$ and $y = -4$.

19. Evaluate $|x + y| - |x|$ for $x = -5$ and $y = 12$.

20. Simplify $14x + 21 - 8x - 49$ by combining like terms.

21. Multiply: $-30(2x + 8)$

22. Find the perimeter of a square with a side length of 19 feet.

Solve the following equations.

23. $11x + 3x = 28$

24. $32 = 7x - 15x$

25. $36x + 11 - 35x - 50 = 10$

26. $6(x - 3) = 0$

27. $9 + 5(3y - 2) = 29$

28. $8(2x - 6) = 4(3x + 5)$

Solve.

29. The sum of five times a number and eight times the same number is thirty-nine. Find the number.

30. Find the length of the side of a square flower bed with a 84-foot perimeter.

Perform the indicated operations and write the answers in lowest terms.

31. $\dfrac{7}{9} \cdot \dfrac{54}{49}$

32. $\dfrac{7x}{5} - \dfrac{2x}{5}$

33. $\dfrac{x^2y}{z} \cdot \dfrac{z^2}{x^3y^2}$

34. $\dfrac{8y}{17} + \dfrac{4}{51}$

35. $2\dfrac{1}{8} \div \dfrac{3}{4}$

36. $7\dfrac{1}{2} + 4\dfrac{3}{8} + 9\dfrac{5}{6}$

37. $21 \div 6\dfrac{3}{7}$

38. $10y^3 \div \dfrac{y}{4}$

39. $-\dfrac{6}{7} \div \left(-\dfrac{36}{35}\right)$

40. $4 \div \left(\dfrac{2}{3} \cdot \dfrac{9}{16}\right)$

41. Simplify $\dfrac{\dfrac{2}{3} + \dfrac{3}{4}}{\dfrac{1}{6} - \dfrac{1}{12}}$.

Solve.

42. $\dfrac{x}{6} + x = -\dfrac{5}{12}$

43. $\dfrac{3}{8} + \dfrac{x}{16} = \dfrac{1}{4} - \dfrac{x}{2}$

44. Evaluate $\dfrac{x}{y}$ for $x = -\dfrac{3}{7}$ and $y = 1\dfrac{1}{9}$.

Perform the indicated operation. Round the result to the nearest thousandth if necessary.

45. $8.23 + 5.607 + 1.9$

46. $-18.6 - 31.5$

47. $(2.7)(20.38)$

48. $(-2.3)^2 + 4.1$

49. $\dfrac{0.6 - 2.34}{-0.3}$

50. Round 29.0572 to the nearest thousandth.

51. Write 0.68 as a fraction in simplest form.

52. Write $\dfrac{107}{500}$ as a decimal.

53. Simplify $-7\sqrt{64} - \sqrt{100}$.

54. Find the volume of a cube with edges $1\dfrac{3}{7}$ inches.

55. Find the area of a rectangle with length 2.3 feet and width 4.9 feet.

Complete and graph the ordered-pair solutions of the given equation.

56. $4x - 7y = 28$; (0,), (, 0), (-1,)

57. $y = 2x - 3$; (0,), (, 0), (2,)

Graph each linear equation.

58. $x + 2y = 4$ 59. $x + 4 = 0$ 60. $2y - 3 = 0$ 61. $5x + 6y = -30$

62. Find the mean of the following list of numbers: 108, 112, 137, 142, 98, 156, 110, 118

63. Find the median of the following list of numbers: 12, 61, 58, 32, 14, 16, 80, 57, 49, 36, 17, 25

64. Find the mode of the following list of numbers: 37, 10, 91, 52, 10, 37, 52, 28, 51, 29, 10, 52, 49, 52

65. Find the grade point average. If necessary, round to the nearest hundredth.

Grade	Credit Hours
B	3
A	3
C	4
C	3
B	5

66. Write the ratio as a fraction in simplest form: 62 marbles to 248 marbles.

67. Write 550 miles to 25 gallons as a unit rate.

Solve the proportion for the variable.

68. $\dfrac{9}{60} = \dfrac{x}{20}$ 69. $\dfrac{5.2}{x} = \dfrac{20}{130}$

Convert as indicated.

70. 1092 in. to feet   71. 58 qt to gallons

72. 12000 mg to grams   73. 16 L to milliliters

74. 20°C to degrees Fahrenheit.

Perform the indicated operations.

75. 19 lb 3 oz − 5 lb 12 oz

77. 2 ft 3 in × 8   76. 3.8 kg + 160 g

79. Write 0.21% as a decimal.   78. Write 0.126 as a percent.

81. What is 5.4% of 1200?   80. Write $\dfrac{6}{75}$ as a percent.

83. 28.5 is 30% of what number?

82. 172.2 is what percent of 820?

84. Find the simple interest earned on $1600 saved for 2½ years at an interest rate of 8.75%.

85. $7000 is borrowed for 4 years. If the interest on the loan is $1248, find the monthly payment.

86. Suzanna is paid a commission of 4.5% of her sales. Find her commission for a week in which her sales were $283,000.

Add or subtract as indicated.

87. $(8x^2 - 6x + 1) + (5x^2 - 4x - 10)$

89. Find the value of $2x^2 + x - 9$ for $x = 3$.

88. $(9x^3 - x + 5) - (4x^2 - 7x + 6)$

Multiply and simplify.

90. $y^{16} \cdot y^{21}$

91. $(x^{16})^4$

92. $(b^7)^3(b^3)^7$

93. $8a(a^2 + 4a - 10)$

94. $(x + 8)(x + 3)$

95. $(x - 7)(x + 7)$

96. $(3y - 1)^2$

97. $(b + 10)(b^2 - 10b + 100)$

98. Find the greatest common factor of $21y^5$, $49y^4$, $98y^{10}$

Factor.

99. $15x^6 - 45x^2$

100. $10a^4 - 30a^3 + -20a^2$

1.1 SOLUTIONS TO EXERCISES

1. 2418 hundreds

2. 45,691 ten-thousands

3. 435,019 hundred thousands

4. 47,912,030 ten-millions

5.
number in period	name of period	number in period
seven	thousand,	three hundred ninety-one

6.
number in period	name of period	number in period
twenty	thousand,	sixty-eight

7.
number in period	name of period	number in period

One hundred twenty-seven thousand, four hundred fifty-five

8.
number in period	name of period	number in period
five	thousand,	forty

9. 5031

10. 8,241,007

11. 28,004

12. 2,000,005,009

13. 500 + 30 + 1

14. 7000 + 800 + 9

15. 60,000,000 + 1,000,000 + 700,000 + 30,000 + 8000 + 400 + 20 + 5

16. 40,000,000 + 50,000 + 60 + 1

17. 9 < 21

18. 10 > 2

19. 31 > 29

20. 51 < 61

21. 4145: four thousand, one hundred forty-five miles

22. 3964: 3000 + 900 + 60 + 4

1.2 SOLUTIONS TO EXERCISES

1.
$$\begin{array}{r} \overset{1}{19} \\ +\ 36 \\ \hline 55 \end{array}$$

2.
$$\begin{array}{r} 81 \\ +\ 75 \\ \hline 156 \end{array}$$

3.
$$\begin{array}{r} \overset{1}{123} \\ +\ 985 \\ \hline 1108 \end{array}$$

4.
$$\begin{array}{r} \overset{1}{27} \\ +\ 47 \\ \hline 74 \end{array}$$

5.
$$\begin{array}{r} 50 \\ +\ 49 \\ \hline 99 \end{array}$$

6.
$$\begin{array}{r} \overset{11}{378} \\ +\ 124 \\ \hline 502 \end{array}$$

7.
$$\begin{array}{r} \overset{11}{3246} \\ +\ 289 \\ \hline 3535 \end{array}$$

118

8.
$$\begin{array}{r} {}^{1} \\ 5177 \\ +\ 4960 \\ \hline 10{,}137 \end{array}$$

9.
$$\begin{array}{r} {}^{1} \\ 50 \\ 47 \\ +\ 68 \\ \hline 165 \end{array}$$

10.
$$\begin{array}{r} {}^{2} \\ 18 \\ {}^{2} \\ 134 \\ 298 \\ +\ 1357 \\ \hline 1807 \end{array}$$

11.
$$\begin{array}{r} {}^{2} \\ 19 \\ 31 \\ 18 \\ 32 \\ +\ 61 \\ \hline 161 \end{array}$$

12.
$$\begin{array}{r} {}^{1} \\ 1972 \\ +\ 8314 \\ \hline 10{,}286 \end{array}$$

13.
$$\begin{array}{r} {}^{1\ 1\ \ 1\ 1} \\ 12{,}388 \\ +\ 9{,}615 \\ \hline 22{,}003 \end{array}$$

14.
$$\begin{array}{r} {}^{1} \\ {}_{1}22 \\ 378 \\ +\ 95{,}416 \\ \hline 95{,}816 \end{array}$$

15.
$$\begin{array}{r} 4000 \\ 200 \\ +\ 6091 \\ \hline 10{,}291 \end{array}$$

16.
$$\begin{array}{r} {}^{1\ 2\ 1\ \ 1\ 1} \\ 321{,}475 \\ 109{,}316 \\ 25{,}492 \\ +\ 87{,}500 \\ \hline 543{,}783 \end{array}$$

17. Add the lengths of the sides:
3 feet + 7 feet + 3 feet + 7 feet = 20 feet

18. Add the lengths of the sides:
5 inches + 2 inches + 5 inches + 4 inches
+ 8 inches = 24 inches

19. original price + mark-up = total price
          189   +   36   =   225

The suit sells for $225.

20. Andrea's money + Kelsey's money = total money
         47   +   18   = $65

65 > 62
They have enough money to buy the game.

1.3 SOLUTIONS TO EXERCISES

1.
$$\begin{array}{r} 59 \\ -\ 34 \\ \hline 25 \end{array} \qquad \text{Check:} \qquad \begin{array}{r} 25 \\ +\ 34 \\ \hline 59 \end{array}$$

2.
$$\begin{array}{r} 65 \\ -\ 25 \\ \hline 40 \end{array} \qquad \text{Check:} \qquad \begin{array}{r} 40 \\ +\ 25 \\ \hline 65 \end{array}$$

3.
$$\begin{array}{r} 873 \\ -\ 642 \\ \hline 231 \end{array}$$
Check:
$$\begin{array}{r} 231 \\ +\ 642 \\ \hline 873 \end{array}$$

4.
$$\begin{array}{r} {\scriptstyle 5\ 11} \\ 6\!\!\not1\!3 \\ -\ 292 \\ \hline 321 \end{array}$$
Check:
$$\begin{array}{r} {\scriptstyle 1} \\ 321 \\ +\ 292 \\ \hline 613 \end{array}$$

5.
$$\begin{array}{r} {\scriptstyle 7\ 11} \\ 8\!\!\not1 \\ -\ 69 \\ \hline 12 \end{array}$$
Check:
$$\begin{array}{r} {\scriptstyle 1} \\ 12 \\ +\ 69 \\ \hline 81 \end{array}$$

6.
$$\begin{array}{r} {\scriptstyle 4\ 13} \\ 5\!\!\not3 \\ -\ 47 \\ \hline 6 \end{array}$$
Check:
$$\begin{array}{r} {\scriptstyle 1} 6 \\ +\ 47 \\ \hline 53 \end{array}$$

7.
$$\begin{array}{r} 299 \\ -\ 187 \\ \hline 112 \end{array}$$
Check:
$$\begin{array}{r} 112 \\ +\ 187 \\ \hline 299 \end{array}$$

8.
$$\begin{array}{r} {\scriptstyle 9} \\ {\scriptstyle 4\ \not1 010} \\ 5\!\!\not0\!\!\not0 \\ -\ 199 \\ \hline 301 \end{array}$$
Check:
$$\begin{array}{r} {\scriptstyle 1\ 1} \\ 301 \\ +\ 199 \\ \hline 500 \end{array}$$

9.
$$\begin{array}{r} {\scriptstyle 9} \\ {\scriptstyle 3\ \not0 11} \\ 4\!\!\not0\!1 \\ -\ 58 \\ \hline 343 \end{array}$$
Check:
$$\begin{array}{r} {\scriptstyle 1\ 1} \\ 343 \\ +\ 58 \\ \hline 401 \end{array}$$

10.
$$\begin{array}{r} {\scriptstyle 4\ 11} \\ 5\!\!\not1 \\ -\ 37 \\ \hline 14 \end{array}$$
Check:
$$\begin{array}{r} {\scriptstyle 1} \\ 14 \\ +\ 37 \\ \hline 51 \end{array}$$

11.
$$\begin{array}{r} 28 \\ -\ 16 \\ \hline 12 \end{array}$$
Check:
$$\begin{array}{r} 12 \\ +\ 16 \\ \hline 28 \end{array}$$

12.
$$\begin{array}{r} {\scriptstyle 11} \\ {\scriptstyle 0\ \not1 211} \\ 121 \\ -\ 93 \\ \hline 28 \end{array}$$
Check:
$$\begin{array}{r} {\scriptstyle 1} \\ 28 \\ +\ 93 \\ \hline 121 \end{array}$$

13.
$$\begin{array}{r} {\scriptstyle 10} \\ {\scriptstyle 4\ \not1 17} \\ 5\!\!\not1\!7 \\ -\ 468 \\ \hline 49 \end{array}$$
Check:
$$\begin{array}{r} {\scriptstyle 1} \\ {\scriptstyle 1}\ 49 \\ +\ 468 \\ \hline 517 \end{array}$$

14.
$$\begin{array}{r} {\scriptstyle 9} \\ {\scriptstyle 1\ \not0 17} \\ 207 \\ -\ 99 \\ \hline 108 \end{array}$$
Check:
$$\begin{array}{r} {\scriptstyle 1\ 1} \\ 108 \\ +\ 99 \\ \hline 207 \end{array}$$

15.
$$\begin{array}{r} {\scriptstyle 1\ 13\ 0\ 17} \\ 2\!\!\not3\!\!\not1\!7 \\ -\ 1809 \\ \hline 508 \end{array}$$
Check:
$$\begin{array}{r} {\scriptstyle 1} \\ {\scriptstyle 1}\ 508 \\ +\ 1809 \\ \hline 2317 \end{array}$$

16.
$$\begin{array}{r} 35 \\ -\ 13 \\ \hline 22 \end{array}$$
Check:
$$\begin{array}{r} 22 \\ +\ 13 \\ \hline 35 \end{array}$$

17.
$$\begin{array}{r} {\scriptstyle 2\ 11} \\ 3\!\!\not1 \\ -\ 16 \\ \hline 15 \end{array}$$
Check:
$$\begin{array}{r} {\scriptstyle 1} \\ 15 \\ +\ 16 \\ \hline 31 \end{array}$$

18. ending   –   beginning   =   miles
     reading   reading   driven

     63,271   –   61,894   =   miles driven

$$\begin{array}{r} \scriptstyle 1116 \\ \scriptstyle 2\not2\not7 11 \\ 63{,}271 \\ -\ 61{,}894 \\ \hline 1{,}377 \end{array}$$

She drove 1,377 miles.

19. normal   –   discount   =   sale
     price   price

     799   –   87   =   sale price

$$\begin{array}{r} 799 \\ -\ 87 \\ \hline 712 \end{array}$$

The sale price is $712.

20. temperature   –   degrees   =   temperature
     at 5 p.m.   dropped   at 8 p.m.

     81   –   13   =   temperature at 8 p.m.

$$\begin{array}{r} \scriptstyle 7\ 11 \\ 8\!\!\!/1 \\ -\ 13 \\ \hline 68 \end{array}$$

The temperature at 8 p.m. was 68°.

1.4 SOLUTIONS TO EXERCISES

1. 5$\underline{9}$1: Since 9 ≥ 5, add 1 to the 5 in the
hundreds place and replace each digit
to the right by 0.

     600

2. 37$\underline{2}$: Since 2 < 5, do not add 1 to the 7 in
the tens place, but replace each digit
to the right by 0.

     370

3. 6$\underline{8}$7: Since 8 ≥ 5, add 1 to the 6 in the
hundreds place and replace each digit
to the right by 0.

     700

4. 3$\underline{9}$18: Since 9 ≥ 5, add 1 to the 3 in the
thousands place and replace each digit
to the right by 0.

     4000

5. 8$\underline{9}$13: Since 9 ≥ 5, add 1 to the 8 in the
thousands place and replace each digit
to the right by 0.

     9000

6. 19$\underline{7}$4: Since 7 ≥ 5, add 1 to the 9 in the
hundreds place and replace each digit
to the right by 0.

     2000

7. 16,3$\underline{9}$9: Since 9 ≥ 5, add 1 to the 3 in the
hundreds place and replace each digit
to the right by 0.

     16,400

8. 21,$\underline{5}$89: Since 5 ≥ 5, add 1 to the 1 in the
thousands place and replace each digit
to the right by 0.

     22,000

9. 73,$\underline{3}$694: Since 3 < 5, do not add 1 to the 7 in
the ten-thousands place, but replace
each digit to the right by 0.

     70,000

10. 49,9$\underline{9}$9: Since 9 ≥ 5, add 1 to the 9 in the
tens place and replace each digit
to the right by 0.

     50,000

11.
```
   38      40
   47      50
   21      20
 + 15    + 20
         130
```

12.
```
   86      90
   74      70
   63      60
 + 99    + 100
         320
```

13.
```
          6 13
  731     7̶3̶0
- 187    - 190
         540
```

14.
```
             1
  918      920
  461      460
+ 379    + 380
         1760
```

15.
```
  604      600
- 399    - 400
         200
```

16.
```
            1 15 17
  2673    2̶6̶7̶0
- 1892   - 1890
         780
```

17.
```
  436     ₁ 400
 1234     1200
+ 765   + 800
         2400
```
They have approximately $2400.

18.
```
 55,674   56,000
- 52,821  - 53,000
          3,000
```

There is an increase of approximately 3000 credit hours.

1.5 SOLUTIONS TO EXERCISES

1. $5(2 + 7) = 5 \cdot 2 + 5 \cdot 7$

2. $8(1 + 4) = 8 \cdot 1 + 8 \cdot 4$

3. $12(20 + 3) = 12 \cdot 20 + 12 \cdot 3$

4.
```
   39
  x 3
  117
```

5.
```
   21
  x 7
  147
```

6.
```
  236
  x 7
 1652
```

7.
```
   243
  x 21
   243
  4860
  5103
```

8.
```
   798
  x 74
  3192
 55860
 59,052
```

9.
```
   2134
   x 62
   4268
 128 040
 132,308
```

10.
```
    869
   x 20
    000
  17 380
  17,380
```

11.
```
    124
  x 231
    124
   3720
  24 800
  28,644
```

12.
```
   2645
 x 237
  18515
  79350
 529000
 626,865
```

13.
```
      80
    x 70
      00
    5600
    5600
```

14. $(19)(1) = 19$
```
     19
   x 30
     00
    570
    570
```

15. $(297)(31)(0)$

= $(297)(0)(31)$

= $(0)(31)$

= 0

16.
```
    483            500
  x 291          x   300
               150,000
```

17.
```
    213            200
  x 365          x   400
                80,000
```

18. Area = width · length
     = 8 inches · 13 inches
     = 104 square inches

19. Area = width · length
     = 3 meters · 12 meters
     = 36 square meters

20.
number of characters per second	·	number of seconds	=	total number of characters
65	·	38	=	total number of characters

```
     65
   x 38
    520
   1950
   2470
```

It can print 2470 characters in 38 seconds.

1.6 SOLUTIONS TO EXERCISES

1.
```
       13
    7)91          Check:    13
      7                   x 7
     21                     91
     21
      0
```

2.
```
       19
    5)95          Check:    19
      5                   x 5
     45                     95
     45
      0
```

3.
```
       38
    4)152         Check:    38
     12                   x 4
     32                    152
     32
      0
```

4.
```
      46 R 4
    5)234         Check:    46
     20                   x 5
     34                    230
     30                   + 4
      4                    234
```

5.   109 R 2
     3)329   Check: 109
        3   × 3
        02   327
         0   + 2
        29   329
        27
         2

6.   47
     21)987   Check: 47
        84   × 21
       147   47
       147   940
         0   987

7.   80 R 1
     4)321   Check: 80
       32   × 4
       01   320
        0   + 1
        1   321

8.   120 R 1
     5)601   Check: 120
       5   × 5
       10   600
       10   + 1
       01   601
        0
        1

9.   500 R 5
     11)5505   Check: 500
        55   × 11
        00   500
         0   5000
        05   5500
         0   + 5
         5   5505

10.   406 R 5
     17)6907   Check: 406
        68   × 17
        10   2842
         0   4060
       107   6902
       102   + 5
         5   6907

11.   101 R 2
     36)3638   Check: 101
        36   × 36
        03   606
         0   3030
        38   3636
        36   + 2
         2   3638

12.   38 R 4
     54)2056   Check: 38
       162   × 54
       436   152
       432   1900
         4   2052
                           + 4
                           2056

13.   34
     192)6528   Check: 192
        576   × 34
        768   768
        768   5760
          0   6528

14.   57
     231)13167   Check: 231
        1155   × 57
        1617   1617
        1617   11550
           0   13,167

15.   111 R 4
     120)13324   Check: 120
        120   × 111
        132   120
        120   1200
        124   12000
        120   13320
          4   + 4
                           13,324

16.   Average = $\dfrac{91 + 86 + 74 + 99 + 115}{5}$

           = $\dfrac{465}{5}$

           = 93

17. Average $= \dfrac{18 + 23 + 14 + 26 + 28 + 31 + 7}{7}$

$\qquad = \dfrac{147}{7}$

$\qquad = 21$

18. Total money $\div$ number of people = money per person
    4,485,000 $\div$ 13   =   345,000

    Each person receives $345,000.

19. number of miles $\cdot$ feet per mile   =   number of feet
         2         $\cdot$ 5280   =   10560

    number of feet $\div$ feet apart   =   number of poles
       10560 $\div$ 483   =   21 R 417

    There are 21 poles.

20. number of miles $\cdot$ number of feet   =   total number of feet
         2         $\cdot$ 5280   =   10,560

    number of feet $\div$ number of feet per yd   =   number of yds
       10,560 $\div$ 3   =   3520

    There are 3520 yards in 2 miles.

1.7 SOLUTIONS TO EXERCISES

1. $5 \cdot 5 \cdot 5 \cdot 5 \cdot 5 \cdot 5 = 5^6$

2. $2 \cdot 2 \cdot 2 \cdot 2 \cdot 2 \cdot 2 \cdot 2 \cdot 2 = 2^8$

3. $8 \cdot 8 \cdot 3 \cdot 3 \cdot 3 = 8^2 \cdot 3^3$

4. $7 \cdot 4 \cdot 4 \cdot 9 \cdot 9 \cdot 9 \cdot 9 \cdot 9 = 7 \cdot 4^2 \cdot 9^5$

5. $7^2 = 7 \cdot 7 = 49$

6. $8^3 = 8 \cdot 8 \cdot 8 = 512$

7. $5^4 = 5 \cdot 5 \cdot 5 \cdot 5 = 625$

8. $2^9 = 2 \cdot 2 \cdot 2 \cdot 2 \cdot 2 \cdot 2 \cdot 2 \cdot 2 \cdot 2 = 512$

9. $3^6 = 3 \cdot 3 \cdot 3 \cdot 3 \cdot 3 \cdot 3 = 729$

10. $7^4 = 7 \cdot 7 \cdot 7 \cdot 7 = 2401$

11. $10^6 = 1,000,000$ (exponent 6: 6 zeros)

12. $9^4 = 9 \cdot 9 \cdot 9 \cdot 9 = 6561$

13. $2180^1 = 2180$

14. $17 + 4 \cdot 8 = 17 + 32$
    $\qquad\qquad\quad = 49$

15. $9 \cdot 5 - 12 = 45 - 12$
    $\qquad\qquad\quad = 33$

16. $0 \div 10 + 8 \cdot 3 = 0 + 24$
    $\qquad\qquad\qquad\quad = 24$

17. $(11 + 5^2) + 6 = (11 + 25) + 6$
    $\qquad\qquad\qquad = 36 \div 6$
    $\qquad\qquad\qquad = 6$

18. $(10 - 3) \cdot (23 - 19) = 7 \cdot 4$
    $\qquad\qquad\qquad\qquad = 28$

19. $(24 \div 8) + [(4 + 2) \cdot 3] = 3 + (6 \cdot 3)$
    $\qquad\qquad\qquad\qquad\qquad = 3 + 18$
    $\qquad\qquad\qquad\qquad\qquad = 21$

20. $42 - \{7 + 4[6 \cdot (12 - 9)] - 60\}$
    $= 42 - \{7 + 4[6 \cdot 3] - 60\}$
    $= 42 - [7 + 4(18) - 60]$
    $= 42 - (7 + 72 - 60)$
    $= 42 - (19)$
    $= 23$

1.8 SOLUTIONS TO EXERCISES

1. $4 + 3y = 4 + 3(6)$
   $\qquad\quad = 4 + 18$
   $\qquad\quad = 22$

2. $2xy - 5z = 2(3)(6) - 5(4)$
   $\qquad\qquad = 36 - 20$
   $\qquad\qquad = 16$

3. $z + 3x - y = 4 + 3(3) - 6$
   $\qquad\qquad\quad = 4 + 9 - 6$
   $\qquad\qquad\quad = 7$

4. $2xy^2 + 8 = 2(3)(6)^2 + 8$
   $\qquad\qquad = 2(3)(36) + 8$
   $\qquad\qquad = 216 + 8$
   $\qquad\qquad = 224$

5. $12 - (y - x) = 12 - (6 - 3)$
$= 12 - 3$
$= 9$

6. $19 + (3x - 5) = 19 + [3(3) - 5]$
$= 19 + (9 - 5)$
$= 19 + 4$
$= 23$

7. $x^3 + z^2 - y = (3)^3 + (4)^2 - 6$
$= 27 + 16 - 6$
$= 37$

8. $z^2 - (x + y) = 4^2 - (3 + 6)$
$= 16 - 9$
$= 7$

9. $\dfrac{2xy}{z} = \dfrac{2(3)(6)}{4}$

$= \dfrac{36}{4}$

$= 9$

10. $\dfrac{8xyz}{9} = \dfrac{8(3)(6)(4)}{9}$

$= \dfrac{576}{9}$

$= 64$

11. $\dfrac{3y + 2}{10} = \dfrac{3(6) + 2}{10}$

$= \dfrac{18 + 2}{10}$

$= \dfrac{20}{10}$

$= 2$

12. $\dfrac{7z}{2} + \dfrac{18}{y} = \dfrac{7(4)}{2} + \dfrac{18}{6}$

$= \dfrac{28}{2} + \dfrac{18}{6}$

$= 14 + 3$
$= 17$

13. $4z(x + y) = 4(4)(3 + 6)$
$= 4(4)(9)$
$= 144$

14. $(xz - 5)^2 = [(3)(4) - 5]^2$
$= (12 - 5)^2$
$= 7^2$
$= 49$

15. sum: addition

words: a number and twelve
translate: x + 12

16. words: a number less twenty-four
translate: x − 24

17. words: a number divided by eight
translate: x ÷ 8

or: $\dfrac{x}{8}$

18. difference: subtraction

words: a number difference one hundred
translate: x − 100

19. words: eight increased by product of four and a number
translate: 8 + $4x$

20. words: difference of eleven added quotient of a
and a number number and four
translate: $(11 - x)$ + $\dfrac{x}{4}$

$(11 - x) + \dfrac{x}{4}$

CHAPTER 1 PRACTICE TEST SOLUTIONS

1.   48
    $\underline{+\ 71}$
       119

2.
$$\begin{array}{r} \overset{9}{4\,\cancel{0}11} \\ \cancel{5}\cancel{0}1 \\ -\;398 \\ \hline 103 \end{array}$$

3.
$$\begin{array}{r} 281 \\ \times\;26 \\ \hline 1686 \\ 5620 \\ \hline 7306 \end{array}$$

4.
$$\begin{array}{r} 577\,R\;61 \\ 72\overline{)41605} \\ \underline{360} \\ 560 \\ \underline{504} \\ 565 \\ \underline{504} \\ 61 \end{array}$$

5. $3^4 \cdot 4^2 = 3 \cdot 3 \cdot 3 \cdot 3 \cdot 4 \cdot 4$
             $= 81 \cdot 16$
             $= 1296$

6. $7^1 \cdot 3^2 = 7 \cdot 3 \cdot 3$
           $= 63$

7. $101 \div 1 = 101$

8. $0 \div 52 = 0$

9. $29 \div 0$ undefined

10. $(7^2 - 6) \cdot 4 = (49 - 6) \cdot 4$
                $= 43 \cdot 4$
                $= 172$

11. $18 + 10 \div 2 \cdot 3 - 9$
   $= 18 + 5 \cdot 3 - 9$
   $= 18 + 15 - 9$
   $= 24$

12. $3[(7 - 5)^2 + (31 - 27)^2] \div 6$
   $= 3[(2)^2 + (4)^2] \div 6$
   $= 3(4 + 16) \div 6$
   $= 3(20) \div 6$
   $= 60 \div 6$
   $= 10$

13. $428{,}791$: Since $8 \ge 5$, add 1 to the 2 in the ten-thousands place and replace each digit to the right by 0.

     $430{,}000$

14.
$$\begin{array}{r} 5176 \\ 3428 \\ +\;769 \\ \hline \end{array} \qquad \begin{array}{r} 5200 \\ 3400 \\ +\;800 \\ \hline 9400 \end{array}$$

15.
$$\begin{array}{r} 7351 \\ -\;2067 \\ \hline \end{array} \qquad \begin{array}{r} 7400 \\ -\;2100 \\ \hline 5300 \end{array}$$

16. Total cost $\div$ number of cans $=$ cost per can
     713      $\div$ 31 $=$ 23

Each can cost \$23.

17. cost per ticket $\cdot$ number of members $=$ total cost
     18         $\cdot$ 12 $=$ 216

The total cost is \$216.

18. higher price $-$ lower price $=$ difference in price
     641         $-$ 489 $=$ 152

The higher-priced stove is \$152 more than the other stove.

19. $2(x^4 - 7) = 2[(3)^4 - 7]$
             $= 2(81 - 7)$
             $= 2(74)$
             $= 148$

20. $\dfrac{4x - 18}{3y} = \dfrac{4(9) - 18}{3(2)}$

           $= \dfrac{36 - 18}{6}$

           $= \dfrac{18}{6}$

           $= 3$

21. (a) difference : subtraction

    words: a number   difference   28
    translate: x –   28

    (b) Three   more than   twice a number
         3   + $2x$
        $2x$ +   3

22. Square:

    Perimeter = 4 · side
               = 4 (4 feet)
               = 16 feet

    Area = $(\text{side})^2$
          = $(4 \text{ feet})^2$
          = 16 square feet

23. Rectangle:

    Perimeter: Add the lengths of the sides
                  14 cm + 5 cm + 14 cm + 5 cm
                  = 38 centimeters

    Area = width · length
          = (5 cm) · (14 cm)
          = 70 square centimeters

24. 546 (Montana row, 1998 Math column)

25. 1999 score   –   1997 score   =   increase
       586   –   582   =   4

    There was an increase of 4 points.

2.1 SOLUTIONS TO EXERCISES

1. −18

2. +75

3.

4.

5. −8 < 3, since −8 is to the left of 3 on the number line.

6. −21 > −22, since −21 is to the right of −22 on the number line.

7. 0 > −4, since 0 is to the right of −4 on the number line.

8. −19 < 19, since −19 is to the left of 19 on the number line.

9. $|-98| = 98$, because −98 is 98 units from 0.

10. $|22| = 22$, because 22 is 22 units from 0.

11. $|-16| = 16$, because −16 is 16 units from 0.

12. −17

13. $-(-30) = 30$

14. −101

15. $-|28| = -28$

16. $-|-17| = -(17) = -17$

17. $-(-44) = 44$

18. $-|-11|$ $-|-10|$
    −11   −10
    −11   <   −10

19. $-|26|$ $-|-26|$
    −26   −26
            −26 = −26

20. −51 $-(-54)$
    −51   54
        −51 < 54

2.2 SOLUTIONS TO EXERCISES

1. $21 + 14 = 35$

2. $-8 + (-1) = -9$

3. $61 + (-61) = 0$

4. $13 + (-6) = 7$

5. $17 + (-22) = -5$

6. $-14 + (-31) = -45$

7. $-14 + 5 + (-8) = -9 + (-8)$
    $= -17$

8. $16 + (-4) + (-5) = 12 + (-5)$
    $= 7$

9. $19 + 7 + (-6) + (-22) = 26 + (-6) + (-22)$
    $= 20 + (-22)$
    $= -2$

10. $(-13) + 12 + (-11) + 10 = -1 + (-11) + 10$
     $= -12 + 10$
     $= -2$

11. $-99 + (-81) = -180$

12. $117 + (-54) = 63$

13. $-87 + 19 = -68$

14. $(-6) + (-10) + 18 + (-2) = -16 + 18 + (-2)$
     $= 2 + (-2)$
     $= 0$

15. $x + y = -15 + 14$
     $= -1$

16. $x + y = -60 + (-55)$
          $= -115$

17. $x + y = 21 + (-33)$
          $= -12$

18. $x + y = -2 + (-82)$
          $= -84$

19. temperature   +   degrees   =   temperature
   at 3:00 p.m.   dropped   at 10:30 p.m.

      -12 + (-8) = -20

   At 10:30 p.m. the temperature was $-20°$ C.

20. depth of original   +   additional   =   new
       dive   feet   depth

      -175 + (-18) = -193

   His new depth is 193 feet below the surface.

2.3 SOLUTIONS TO EXERCISES

1. $-17 - (-17) = -17 + 17$
                 $= 0$

2. $8 - 5 = 8 + (-5)$
         $= 3$

3. $11 - 16 = 11 + (-16)$
            $= -5$

4. $-7 - (-9) = -7 + 9$
             $= 2$

5. $9 - 10 = 9 + (-10)$
          $= -1$

6. $-12 - 52 = -12 + (-52)$
             $= -64$

7. $-33 - 20 = -33 + (-20)$
             $= -53$

8. $-40 - (-45) = -40 + 45$
              $= 5$

9. $42 - 35 - 7 = 42 + (-35) + (-7)$
              $= 7 + (-7)$
              $= 0$

10. $-8 - 9 - (-12) = -8 + (-9) + 12$
                  $= -17 + 12$
                  $= -5$

11. $14 - (-16) + 8 = 14 + 16 + 8$
                 $= 30 + 8$
                 $= 38$

12. $-(-13) - 17 + 8 = 13 + (-17) + 8$
                  $= -4 + 8$
                  $= 4$

13. $-20 - 20 - 4 = -20 + (-20) + (-4)$
                 $= -40 + (-4)$
                 $= -44$

14. $-20 - (-20) - 4 = -20 + 20 + (-4)$
                  $= 0 + (-4)$
                  $= -4$

15. $x - y = -17 - 11$
          $= -17 + (-11)$
          $= -28$

16. $x - y = -39 - (-68)$
          $= -39 + 68$
          $= 29$

17. $x - y + z = -7 - 6 + 12$
             $= -7 + (-6) + 12$
             $= -13 + 12$
             $= -1$

18. $-x - y + z = -40 - (-50) + (-60)$
              $= -40 + 50 + (-60)$
              $= 10 + (-60)$
              $= -50$

19.

beginning balance	–	amount of check	+	amount of deposit	–	amount of check	=	current balance
250	–	98	+	39	–	111	=	current balance

$$250 - 98 + 39 - 111 = 250 + (-98) + 39 + (-111)$$
$$= 152 + 39 + (-111)$$
$$= 191 + (-111)$$
$$= 80$$

His current balance is $80.

20.

current score	–	points lost	=	new score
20	–	30	=	new score

$$20 - 30 = 20 + (-30)$$
$$= -10$$

Omar has a new score of -10.

2.4 SOLUTIONS TO EXERCISES

1. $(-8)(-11) = 88$

2. $0(-23) = 0$

3. $(-21)(2) = -42$

4. $\dfrac{-20}{-5} = 4$

5. $\dfrac{54}{-6} = -9$

6. $\dfrac{101}{0}$ is undefined

7. $(4)(-3)(-2) = (-12)(-2)$
   $$= 24$$

8. $(-6)(-10)(-3) = (60)(-3)$
   $$= -180$$

9. $(-1)(5)(-4)(-1) = (-5)(-4)(-1)$
   $$= (20)(-1)$$
   $$= -20$$

10. $(-4)^3 = (-4)(-4)(-4)$
    $$= (16)(-4)$$
    $$= -64$$

11. $(-8)^2 = (-8)(-8)$
    $$= 64$$

12. $(-39)(13) = -507$

13. $a \cdot b = 4 \cdot (-4)$
    $$= -16$$

14. $a \cdot b = (-20) \cdot (-6)$
    $$= 120$$

15. $a \cdot b = 80 \cdot (-3)$
    $$= -240$$

16. $\dfrac{x}{y} = \dfrac{78}{-39}$
    $$= -2$$

17. $\dfrac{x}{y} = \dfrac{-121}{-11}$
    $$= 11$$

18. $\dfrac{x}{y} = \dfrac{0}{99}$
    $$= 0$$

19.

yards lost per play	·	number of plays	=	total loss
-6	·	3	=	-18

Their total loss was 18 yards.

20.
amount lost per day	·	number of days	=	total loss
-200	·	6	=	-1200

He lost $1200.

2.5 SOLUTIONS TO EXERCISES

1. $4 + (-15) \div 3 = 4 + (-5)$
$$= -1$$

2. $9 + 5(4) = 9 + 20$
$$= 29$$

3. $11(-3) + 8 = -33 + 8$
$$= -25$$

4. $15 + 4^3 = 15 + 64$
$$= 79$$

5. $\dfrac{17 - 8}{-9} = \dfrac{9}{-9}$
$$= -1$$

6. $\dfrac{77}{-5 + 12} = \dfrac{77}{7}$
$$= 11$$

7. $[5 + (-3)]^4 = (2)^4$
$$= 16$$

8. $50 - (-6)^2 = 50 - 36$
$$= 14$$

9. $|7 + 4| \cdot (-8)^2 = |11| \cdot (-8)^2$
$$= 11 \cdot (-8)^2$$
$$= 11 \cdot 64$$
$$= 704$$

10. $6 \cdot 3^2 + 15 = 6 \cdot 9 + 15$
$$= 54 + 15$$
$$= 69$$

11. $9 + 3^2 - 4^3 = 9 + 9 - 64$
$$= 18 - 64$$
$$= -46$$

12. $(8 - 16) \div 4 = (-8) \div 4$
$$= -2$$

13. $(24 - 28) \div (31 - 35) = (-4) \div (-4)$
$$= 1$$

14. $(-30 - 6) \div 12 - 11 = -36 \div 12 - 11$
$$= -3 - 11$$
$$= -14$$

15. $-4^2 - (-9)^2 = -16 - 81$
$$= -97$$

16. $2(-15) \div [4(-7) - 9(-3)]$
$$= 2(-15) \div [-28 - (-27)]$$
$$= 2(-15) \div (-1)$$
$$= -30 \div (-1)$$
$$= 30$$

17. $\dfrac{(-5)(-6) - (7)(6)}{2[9 \div (10 - 13)]} = \dfrac{30 - 42}{2[9 \div (-3)]}$
$$= \dfrac{-12}{2(-3)}$$
$$= \dfrac{-12}{-6}$$
$$= 2$$

18. $\dfrac{40(-1) - (-4)(-5)}{3[-10 \div (-7 + 2)]} = \dfrac{-40 - 20}{3[-10 \div (-5)]}$
$$= \dfrac{-60}{3(2)}$$
$$= \dfrac{-60}{6}$$
$$= -10$$

19. $3x - y^2 = 3(-3) - 5^2$
$$= 3(-3) - 25$$
$$= -9 - 25$$
$$= -34$$

20. $\dfrac{10x}{z} - 4y = \dfrac{10(-3)}{-1} - 4(5)$

$= \dfrac{-30}{-1} - 20$

$= 30 - 20$
$= 10$

CHAPTER 2 PRACTICE TEST SOLUTIONS

1. $17 - 36 = 17 + (-36)$
$\qquad\quad = -19$

2. $-15 + 9 = -6$

3. $4 \cdot (-25) = -100$

4. $(-18) \div (-6) = 3$

5. $(-28) + (-14) = -42$

6. $-11 - (-21) = -11 + 21$
$\qquad\qquad\quad = 10$

7. $(-20) \cdot (-4) = 80$

8. $\dfrac{-108}{-12} = 9$

9. $|-52| + (-17) = 52 + (-17)$
$\qquad\qquad\quad = 35$

10. $35 - |-100| = 35 - 100$
$\qquad\qquad\quad = -65$

11. $|7| \cdot |-8| = 7 \cdot 8$
$\qquad\qquad = 56$

12. $\dfrac{|-15|}{-|-3|} = \dfrac{15}{-3}$

$\qquad = -5$

13. $(-12) + 8 \div (-4) = (-12) + (-2)$
$\qquad\qquad\qquad\quad = -14$

14. $-6 + (-51) - 13 + 7 = -57 - 13 + 7$
$\qquad\qquad\qquad\qquad\ = -70 + 7$
$\qquad\qquad\qquad\qquad\ = -63$

15. $(-4)^3 - 42 \div (-7) = -64 - 42 \div (-7)$
$\qquad\qquad\qquad\qquad = -64 - (-6)$
$\qquad\qquad\qquad\qquad = -58$

16. $(3 - 10)^2 \cdot (9 - 11)^3 = (-7)^2 \cdot (-2)^3$
$\qquad\qquad\qquad\qquad\qquad = 49 \cdot (-8)$
$\qquad\qquad\qquad\qquad\qquad = -392$

17. $-(-6)^2 \div 12 \cdot (-5) = -36 \div 12 \cdot (-5)$
$\qquad\qquad\qquad\qquad\ = -3 \cdot (-5)$
$\qquad\qquad\qquad\qquad\ = 15$

18. $24 - (18 - 16)^3 = 24 - 2^3$
$\qquad\qquad\qquad\ = 24 - 8$
$\qquad\qquad\qquad\ = 16$

19. $-9 + (-39) \div (-13) = -9 + 3$
$\qquad\qquad\qquad\qquad = -6$

20. $\dfrac{6}{3} - \dfrac{9^2}{27} = \dfrac{6}{3} - \dfrac{81}{27}$

$\qquad\quad = 2 - 3$
$\qquad\quad = -1$

21. $\dfrac{-7(-3) + 13}{-1(-11 - 6)} = \dfrac{21 + 13}{-1(-17)}$

$\qquad\qquad = \dfrac{34}{17}$

$\qquad\qquad = 2$

22. $\dfrac{|45 - 50|^2}{8(-3) + 19} = \dfrac{|-5|^2}{-24 + 19}$

$\qquad\qquad = \dfrac{5^2}{-5}$

$\qquad\qquad = \dfrac{25}{-5}$

$\qquad\qquad = -5$

23. $5x - y = 5(0) - (-4)$
$$= 0 - (-4)$$
$$= 4$$

24. $|x| - |y| - |z|$
$$= |0| - |-4| - |3|$$
$$= 0 - 4 - 3$$
$$= -7$$

25. $\dfrac{12z}{-3y} = \dfrac{12(3)}{-3(-4)}$
$$= \dfrac{36}{12}$$
$$= 3$$

26. $9 - y = 9 - (-4)$
$$= 13$$

27. beginning   +   change in   =   final
     elevation   elevation   elevation

     15,947   +   (−6161)   =   9786

His final elevation is 9786 feet.

28. beginning − 1ˢᵗ check − 2ⁿᵈ check + deposit = ending
     balance   amount   amount   balance

     318   −   219   −   39   +   72   =   132

His ending balance is $132.

3.1 SOLUTIONS TO EXERCISES

1. The numerical coefficient of $22d$ is 22.

2. The numerical coefficient of $-y$ is -1.

3. The numerical coefficient of $14x^3y^2$ is 14.

4. $13x - 4x = (13 - 4)x$
   $\qquad\quad = 9x$

5. $3b - 7b - b = (3 - 7 - 1)b$
   $\qquad\qquad\quad = -5b$

6. $12y + 9y - 2y + 7 = (12 + 9 - 2)y + 7$
   $\qquad\qquad\qquad\quad = 19y + 7$

7. $-5(11x) = (-5 \cdot 11)x$
   $\qquad\quad\; = -55x$

8. $2(3x - 1) = 2 \cdot 3x + 2(-1)$
   $\qquad\qquad = 6x - 2$

9. $7(4x + 8) = 7 \cdot 4x + 7 \cdot 8$
   $\qquad\qquad = 28x + 56$

10. $4(x - 11) + 16 = 4(x) + 4(-11) + 16$
    $\qquad\qquad\qquad = 4x - 44 + 16$
    $\qquad\qquad\qquad = 4x - 28$

11. $2(9 - 3a) - 16a = 2(9) + 2(-3a) - 16a$
    $\qquad\qquad\qquad\; = 18 - 6a - 16a$
    $\qquad\qquad\qquad\; = 18 + (-6 - 16)a$
    $\qquad\qquad\qquad\; = 18 - 22a$

12. $-5(2x + 7) + 4(x - 6)$
    $= -5(2x) + (-5)(7) + 4(x) + 4(-6)$
    $= -10x - 35 + 4x - 24$
    $= -10x + 4x - 35 - 24$
    $= (-10 + 4)x - 35 - 24$
    $= -6x - 59$

13. $-x + 13y - 9x + 18y$
    $= -x - 9x + 13y + 18y$
    $= (-1 - 9)x + (13 + 18)y$
    $= -10x + 31y$

14. $-12(1 + 2v) + 4v$
    $= -12(1) + (-12)(2v) + 4v$
    $= -12 - 24v + 4v$
    $= -12 + (-24 + 4)v$
    $= -12 - 20v$

15. $-4(m - 9) + 8(m + 3)$
    $= -4(m) + (-4)(-9) + 8(m) + 8(3)$
    $= -4m + 36 + 8m + 24$
    $= -4m + 8m + 36 + 24$
    $= (-4 + 8)m + 36 + 24$
    $= 4m + 60$

16. $22x + 6(2x - 5)$
    $= 22x + 6(2x) + 6(-5)$
    $= 22x + 12x - 30$
    $= (22 + 12)x - 30$
    $= 34x - 30$

17. $9(4xy + 3) - 1(2xy - 8)$
    $= 9(4xy) + 9(3) + (-1)(2xy) + (-1)(-8)$
    $= 36xy + 27 - 2xy + 8$
    $= 36xy - 2xy + 27 + 8$
    $= (36 - 2)xy + 27 + 8$
    $= 34xy + 35$

18. Perimeter
    $= $ sum of the side lengths
    $= 2x + x + (5x + 6) + 4x + (7x + 6) + 5x$
    $= 2x + x + 5x + 4x + 7x + 5x + 6 + 6$
    $= 24x + 12$

    The perimeter is $(24x + 12)$ inches.

19. Area of a square $= (\text{side})^2$
    $\qquad\qquad\qquad\; = (6y)^2$
    $\qquad\qquad\qquad\; = 6y \cdot 6y$
    $\qquad\qquad\qquad\; = 36y^2$

    The area is $(36y^2)$ square meters.

20. Total Area = Area of   +   Area of
                  1st rectangle   2nd rectangle

Area of 1st rectangle = length $\cdot$ width
$$= 9 \cdot (8x - 3)$$
$$= 9(8x) + 9(-3)$$
$$= 72x - 27$$

Area of 2nd rectangle = length $\cdot$ width
$$= 20 \cdot (2x + 1)$$
$$= 20(2x) + 20(1)$$
$$= 40x + 20$$

Total Area $= (72x - 27) + (40x + 20)$
$$= 72x + 40x - 27 + 20$$
$$= 112x - 7$$

The area is $(112x - 7)$ square feet.

3.2 SOLUTIONS TO EXERCISES

1. $y + 9 = 17$
    $8 + 9 = 17$?
    $17 = 17$?

    True

    8 is a solution.

2. $x - 12 = -16$
    $-4 - 12 = -16$?
    $-16 = -16$?

    True

    -4 is a solution.

3. $3(a - 2) = 6$
    $3(0 - 2) = 6$?
    $3(-2) = 6$?
    $-6 = 6$?

    False

    0 is not a solution.

4. $5b = 47 - b$
    $5(-3) = 47 - (-3)$?
    $-15 = 50$?

    False

    -3 is not a solution.

5. $x + 7 = 23$
    $x + 7 - 7 = 23 - 7$
    $x = 16$

6. $t - 14 = 21$
    $t - 14 + 14 = 21 + 14$
    $t = 35$

7. $x - 6 = 2 + 8$
    $x - 6 = 10$
    $x - 6 + 6 = 10 + 6$
    $x = 16$

8. $-9 - 2 = -5 + y$
    $-11 = -5 + y$
    $-11 + 5 = -5 + y + 5$
    $-11 + 5 = -5 + 5 + y$
    $-6 = y$

9. $4a + 3 - 3a = -2 + 12$
    $4a - 3a + 3 = -2 + 12$
    $a + 3 = 10$
    $a + 3 - 3 = 10 - 3$
    $a = 7$

10. $11 - 11 = 9x - 8x$
     $0 = x$

11. $20 + (-17) = 10x + 7 - 9x$
     $20 + (-17) = 10x - 9x + 7$
     $3 = x + 7$
     $3 - 7 = x + 7 - 7$
     $-4 = x$

12. $78 = w + 78$
     $78 - 78 = w + 78 - 78$
     $0 = w$

13. $y + 4 = -14$
     $y + 4 - 4 = -14 - 4$
     $y = -18$

14. $m + 20 = 22$
     $m + 20 - 20 = 22 - 20$
     $m = 2$

15. $-33 + x = -35$
     $-33 + x + 33 = -35 + 33$
     $-33 + 33 + x = -35 + 33$
     $x = -2$

16. $-17n - 10 + 18n = -6$
$-17n + 18n - 10 = -6$
$n - 10 = -6$
$n - 10 + 10 = -6 + 10$
$n = 4$

17. $5y + 9 - 4y = 42$
$5y - 4y + 9 = 42$
$y + 9 = 42$
$y + 9 - 9 = 42 - 9$
$y = 33$

18. $-19x + 13 + 20x = -1 - 5$
$-19x + 20x + 13 = -1 - 5$
$x + 13 = -6$
$x + 13 - 13 = -6 - 13$
$x = -19$

19. $31x + 27 - 30x = -27$
$31x - 30x + 27 = -27$
$x + 27 = -27$
$x + 27 - 27 = -27 - 27$
$x = -54$

20. $z - 82 = -25$
$z - 82 + 82 = -25 + 82$
$z = 57$

3.3 SOLUTIONS TO EXERCISES

1. $7x = 56$

$$\frac{7x}{7} = \frac{56}{7}$$

$$\frac{7}{7} \cdot x = \frac{56}{7}$$

$$x = 8$$

2. $-9z = 99$

$$\frac{-9z}{-9} = \frac{99}{-9}$$

$$\frac{-9}{-9} \cdot z = \frac{99}{-9}$$

$$z = -11$$

3. $-5y = -105$

$$\frac{-5y}{-5} = \frac{-105}{-5}$$

$$\frac{-5}{-5} \cdot y = \frac{-105}{-5}$$

$$y = 21$$

4. $2m = 24$

$$\frac{2m}{2} = \frac{24}{2}$$

$$\frac{2}{2} \cdot m = \frac{24}{2}$$

$$m = 12$$

5. $4b - 9b = -35$
$-5b = -35$

$$\frac{-5b}{-5} = \frac{-35}{-5}$$

$$b = 7$$

6. $18 = 6t - 3t$
$18 = 3t$

$$\frac{18}{3} = \frac{3t}{3}$$

$$6 = t \quad \text{or} \quad t = 6$$

7. $3x - 8x = -10 + (-10)$
$-5x = -20$

$$\frac{-5x}{-5} = \frac{-20}{-5}$$

$$x = 4$$

8. $-12y = 12$

$$\frac{-12y}{-12} = \frac{12}{-12}$$

$$y = -1$$

9. $n + 6n = 77$
$$7n = 77$$
$$\frac{7n}{7} = \frac{77}{7}$$
$$n = 11$$

10. $-8x = -8$
$$\frac{-8x}{-8} = \frac{-8}{-8}$$
$$x = 1$$

11. $y - 9y = 48$
$$-8y = 48$$
$$\frac{-8y}{-8} = \frac{48}{-8}$$
$$y = -6$$

12. $14a - 6a = 32$
$$8a = 32$$
$$\frac{8a}{8} = \frac{32}{8}$$
$$a = 4$$

13. $-64 + 26 = -11x - 8x$
$$-38 = -19x$$
$$\frac{-38}{-19} = \frac{-19x}{-19}$$
$$2 = x \quad \text{or} \quad x = 2$$

14. $13v + 2v = 45$
$$15v = 45$$
$$\frac{15v}{15} = \frac{45}{15}$$
$$v = 3$$

15. $11y - 5y = -6 - 60$
$$6y = -66$$
$$\frac{6y}{6} = \frac{-66}{6}$$
$$y = -11$$

16. $32b - 34b = 82$
$$-2b = 82$$
$$\frac{-2b}{-2} = \frac{82}{-2}$$
$$b = -41$$

17.   Let x = a number

Eight   added   the product of nine
         to   and a number

   8   + $9x$

18.   Let x = a number

Three times   decreased   20
  a number   by

   $3x$ $-$ 20

19.   Let x = a number
quotient: divide
product of a number and -4: $-4x$

$$\frac{55}{-4x}$$

20.   Let x = a number
Twice   the difference of
         a number and -6

   2        $[x - (-6)]$

3.4 SOLUTIONS TO EXERCISES

1. $5x - 30 = 0$
$$5x - 30 + 30 = 0 + 30$$
$$5x = 30$$
$$\frac{5x}{5} = \frac{30}{5}$$
$$x = 6$$

2. $2b - 6 = 8$
$$2b - 6 + 6 = 8 + 6$$
$$2b = 14$$
$$\frac{2b}{2} = \frac{14}{2}$$
$$b = 7$$

3.
$$7y - 4 = 3y + 12$$
$$7y - 4 + 4 = 3y + 12 + 4$$
$$7y = 3y + 16$$
$$7y - 3y = 3y + 16 - 3y$$
$$4y = 16$$
$$\frac{4y}{4} = \frac{16}{4}$$
$$y = 4$$

4.
$$9 - a = 11$$
$$9 - a - 9 = 11 - 9$$
$$-a = 2$$
$$-1a = 2$$
$$\frac{-1a}{-1} = \frac{2}{-1}$$
$$a = -2$$

5.
$$6d + 25 = 31$$
$$6d + 25 - 25 = 31 - 25$$
$$6d = 6$$
$$\frac{6d}{6} = \frac{6}{6}$$
$$d = 1$$

6.
$$8y - 3 = -19$$
$$8y - 3 + 3 = -19 + 3$$
$$8y = -16$$
$$\frac{8y}{8} = \frac{-16}{8}$$
$$y = -2$$

7.
$$4p - 28 = 0$$
$$4p - 28 + 28 = 0 + 28$$
$$4p = 28$$
$$\frac{4p}{4} = \frac{28}{4}$$
$$p = 7$$

8.
$$5t + 2 = 37$$
$$5t + 2 - 2 = 37 - 2$$
$$5t = 35$$
$$\frac{5t}{5} = \frac{35}{5}$$
$$t = 7$$

9.
$$9x + 7 = 6x - 8$$
$$9x + 7 - 7 = 6x - 8 - 7$$
$$9x = 6x - 15$$
$$9x - 6x = 6x - 15 - 6x$$
$$3x = -15$$
$$\frac{3x}{3} = \frac{-15}{3}$$
$$x = -5$$

10.
$$-y - 11 = 5y - 17$$
$$-y - 11 + 17 = 5y - 17 + 17$$
$$-y + 6 = 5y$$
$$-y + 6 + y = 5y + y$$
$$6 = 6y$$
$$\frac{6}{6} = \frac{6y}{6}$$
$$1 = y \quad \text{or} \quad y = 1$$

11.
$$-11x + 1 = -12x + 6$$
$$-11x + 1 - 1 = -12x + 6 - 1$$
$$-11x = -12x + 5$$
$$-11x + 12x = -12x + 5 + 12x$$
$$x = 5$$

12.
$$14 - 5y = 14 + 2y$$
$$14 - 5y + 5y = 14 + 2y + 5y$$
$$14 - 14 = 14 + 7y - 14$$
$$0 = 7y$$
$$\frac{0}{7} = \frac{7y}{7}$$
$$0 = y \quad \text{or} \quad y = 0$$

13.
$$3(m - 2) = m - 10$$
$$3m - 6 = m - 10$$
$$3m - 6 + 6 = m - 10 + 6$$
$$3m = m - 4$$
$$3m - m = m - 4 - m$$
$$2m = -4$$
$$\frac{2m}{2} = \frac{-4}{2}$$
$$m = -2$$

14.
$$7(4c + 2) - 1 = 26c + 5$$
$$28c + 14 - 1 = 26c + 5$$
$$28c + 13 = 26c + 5$$
$$28c + 13 - 13 = 26c + 5 - 13$$
$$28c = 26c - 8$$
$$28c - 26c = 26c - 8 - 26c$$
$$2c = -8$$
$$\frac{2c}{2} = \frac{-8}{2}$$
$$c = -4$$

15.
$$9(4 - x) = 4(x - 4)$$
$$36 - 9x = 4x - 16$$
$$36 - 9x + 9x = 4x - 16 + 9x$$
$$36 = 13x - 16$$
$$36 + 16 = 13x - 16 + 16$$
$$52 = 13x$$
$$\frac{52}{13} = \frac{13x}{13}$$
$$4 = x \quad \text{or} \quad x = 4$$

16.
$$20 + 8(y - 1) = 6y + 32$$
$$20 + 8y - 8 = 6y + 32$$
$$12 + 8y = 6y + 32$$
$$12 + 8y - 12 = 6y + 32 - 12$$
$$8y = 6y + 20$$
$$8y - 6y = 6y + 20 - 6y$$
$$2y = 20$$
$$\frac{2y}{2} = \frac{20}{2}$$
$$y = 10$$

17. The sum of is -11
    -24 and 13

    $-24 + 13 \quad = \quad -11$

18. Four   times   the difference   amounts -52
                of -10 and 3 to

    $4 \quad \cdot \quad (-10 - 3) \quad = \quad -52$

19. Twenty subtracted   equals -35
    from -15

    $-15 - 20 \quad\quad = \quad -35$

20. The quotient of   is equal 15
    150 and twice 5 to

    $\dfrac{150}{2(5)} \quad\quad = \quad 15$

3.5 SOLUTIONS TO EXERCISES

1. In words: A number   added -8 is -12
                    to

    Translate: $x \quad + \quad (-8) = -12$

2. In words: Four   times   a number   yields 44

    Translate: $4 \quad \cdot \quad x \quad = \quad 44$

3. In words: Three   added   twice a   gives -11
                 to   number

    Translate: $3 \quad + \quad 2x \quad = \quad -11$

4. In words: Seven   times   the difference   amounts -14
                   of 6 and a number   to

    Translate: $7 \quad \cdot \quad (6 - x) \quad = \quad -14$

5. Let $x =$ the unknown number

    In words: Five times   added   twelve   is   thirty-seven
            a number   to

    Translate: $5x \quad + \quad 12 \quad = \quad 37$

$$5x + 12 = 37$$
$$5x + 12 - 12 = 37 - 12$$
$$5x = 25$$
$$\frac{5x}{5} = \frac{25}{5}$$
$$x = 5$$

The unknown number is 5.

6. Let x = the unknown number

In words: The product of eight and a number gives fifty-six

Translate: $8x = 56$

$$8x = 56$$
$$\frac{8x}{8} = \frac{56}{8}$$
$$x = 7$$

The unknown number is 7.

7. Let x = the unknown number

In words: A number less nine is twelve

Translate: $x - 9 = 12$

$$x - 9 = 12$$
$$x - 9 + 9 = 12 + 9$$
$$x = 21$$

The unknown number is 21.

8. Let x = the unknown number

In words: Fourteen decreased by some number equals the quotient of twenty and four

Translate: $14 - x = \dfrac{20}{4}$

$$14 - x = \frac{20}{4}$$
$$14 - x = 5$$
$$14 - x - 14 = 5 - 14$$
$$-x = -9$$
$$\frac{-x}{-1} = \frac{-9}{-1}$$
$$x = 9$$

The unknown number is 9.

9. Let x = the unknown number

In words: The sum of four, five and a number is ten

Translate: $4 + 5 + x = 10$

$$4 + 5 + x = 10$$
$$9 + x = 10$$
$$9 + x - 9 = 10 - 9$$
$$x = 1$$

The unknown number is 1.

10. Let x = the unknown number

In words: The product of thirteen and a number is one hundred sixty-nine

Translate: $13x = 169$

$$13x = 169$$
$$\frac{13x}{13} = \frac{169}{13}$$
$$x = 13$$

The unknown number is 13.

11. Let x = the unknown number

In words:	Seventeen	added to	the product of five and some number	amounts to	the product of seven and the number	added to	twenty-five
Translate:	17	+	$5x$	=	$7x$	+	25

$$17 + 5x = 7x + 25$$
$$17 + 5x - 5x = 7x + 25 - 5x$$
$$17 = 2x + 25$$
$$17 + (-25) = 2x + 25 + (-25)$$
$$-8 = 2x$$

$$\frac{-8}{2} = \frac{2x}{2}$$

$$-4 = x$$

The unknown number is -4.

12. Let x = the unknown number

In words:	Seventy-three	less	a number	is equal to	the product of 6 and	the sum of the number and four
Translate:	73	−	x	=	6	$(x + 4)$

$$73 - x = 6(x + 4)$$
$$73 - x = 6x + 24$$
$$73 - x + x = 6x + 24 + x$$
$$73 = 7x + 24$$
$$73 - 24 = 7x + 24 - 24$$
$$49 = 7x$$

$$\frac{49}{7} = \frac{7x}{7}$$

$$7 = x$$

The unknown number is 7.

13. Let x = the number of cards Bob has
Then $3x$ = the number of cards Billy has

In words:	Together	they have	960 cards
Translate:	$x + 3x$	=	960

$$x + 3x = 960$$
$$4x = 960$$

$$\frac{4x}{4} = \frac{960}{4}$$

$$x = 240$$

Billy has $3x = 3(240) = 720$ cards.

14.   Let x = trailer value
     Then $5x$ = boat value

In words: A boat and trailer   are worth   7800

Translate: $x + 5x$ $=$ 7800

$$x + 5x = 7800$$
$$6x = 7800$$

$$\frac{6x}{6} = \frac{7800}{6}$$

$$x = 1300$$
$$5x = 5(1300) = 6500$$

The trailer is worth \$1300 and the boat is worth \$6500.

15.   Let x = amount for accessories
     Then $6x$ = amount for dolls

In words: The doll collection and accessories   sold for   560

Translate: $x + 6x$ $=$ 560

$$x + 6x = 560$$
$$7x = 560$$

$$\frac{7x}{7} = \frac{560}{7}$$

$$x = 80$$

The dolls sold for $6x = 6(\$80) = \$480$.

16.   Let x = speed of Karen's car
     Then $2x$ = speed of Doug's car

In words: Their combined speed   is   90

Translate: $x + 2x$ $=$ 90

$$x + 2x = 90$$
$$3x = 90$$

$$\frac{3x}{3} = \frac{90}{3}$$

$$x = 30$$

Doug's speed is $2x = 2(30) = 60$ miles per hour.

SOLUTIONS TO CHAPTER 3 PRACTICE TEST

1. $4x - 7 - 9x + 12 = 4x + (-9x) + (-7) + 12$
$$= -5x + 5$$

2. $-5(4y - 5) = -5(4y) + (-5)(-5)$
$$= -20y + 25$$

3. $7(2z + 9) - 3z + 21 = 14z + 63 - 3z + 21$
$$= 14z + (-3z) + 63 + 21$$
$$= 11z + 84$$

4. $P = 4(\text{side})$
$$= 4[(5x + 8) \text{ meters}]$$
$$= (20x + 32) \text{ meters}$$

5. Area of 1st rectangle: $A = \text{width} \cdot \text{length}$
$$= (3x \text{ feet})(4 \text{ feet})$$
$$= 12x \text{ square feet}$$

   Area of 2nd rectangle: $A = \text{width} \cdot \text{length}$
$$= (9 \text{ feet})[(4x - 2) \text{ feet}]$$
$$= (36x - 18) \text{ square feet}$$

   Total Area $= 1^{st}$ area $+ 2^{nd}$ area
$$= (12x) \text{ sq. ft} + (36x - 18) \text{ sq. ft}$$
$$= (48x - 18) \text{ square feet}$$

6. $7x + x = -64$
$$8x = -64$$
$$\frac{8x}{8} = \frac{-64}{8}$$
$$x = -8$$

7. $29 = 2x - 31x$
$$29 = -29x$$
$$\frac{29}{-29} = \frac{-29x}{-29}$$
$$-1 = x$$

8. $5b - 7 = 28$
$$5b - 7 + 7 = 28 + 7$$
$$5b = 35$$
$$\frac{5b}{5} = \frac{35}{5}$$
$$b = 7$$

9. $14 + 6z = 50$
$$14 + 6z - 14 = 50 - 14$$
$$6z = 36$$
$$\frac{6z}{6} = \frac{36}{6}$$
$$z = 6$$

10. $11x + 21 - 10x - 18 = 30$
$$x + 3 = 30$$
$$x + 3 - 3 = 30 - 3$$
$$x = 27$$

11. $1 - d + 6d = 31$
$$1 + 5d = 31$$
$$1 + 5d - 1 = 31 - 1$$
$$5d = 30$$
$$\frac{5d}{5} = \frac{30}{5}$$
$$d = 6$$

12. $6x - 9 = -57$
$$6x - 9 + 9 = -57 + 9$$
$$6x = -48$$
$$\frac{6x}{6} = \frac{-48}{6}$$
$$x = -8$$

13. $-9y + 8 = -10$
$$-9y + 8 - 8 = -10 - 8$$
$$-9y = -18$$
$$\frac{-9y}{-9} = \frac{-18}{-9}$$
$$y = 2$$

14. $4(x - 3) = 0$
$$4x - 12 = 0$$
$$4x - 12 + 12 = 0 + 12$$
$$4x = 12$$
$$\frac{4x}{4} = \frac{12}{4}$$
$$x = 3$$

15.
$$6(3 + 8y) = 18$$
$$18 + 48y = 18$$
$$18 + 48y - 18 = 18 - 18$$
$$48y = 0$$

$$\frac{48y}{48} = \frac{0}{48}$$

$$y = 0$$

16.
$$11x - 5 = x + 25$$
$$11x - 5 - x = x + 25 - x$$
$$10x - 5 = 25$$
$$10x - 5 + 5 = 25 + 5$$
$$10x = 30$$

$$\frac{10x}{10} = \frac{30}{10}$$

$$x = 3$$

17.
$$14a - 3 = 5a + 24$$
$$14a - 3 + 3 = 5a + 24 + 3$$
$$14a = 5a + 27$$
$$14a - 5a = 5a + 27 - 5a$$
$$9a = 27$$

$$\frac{9a}{9} = \frac{27}{9}$$

$$a = 3$$

18.
$$4 + 5(2m - 3) = 9$$
$$4 + 10m - 15 = 9$$
$$10m - 11 = 9$$
$$10m - 11 + 11 = 9 + 11$$
$$10m = 20$$

$$\frac{10m}{10} = \frac{20}{10}$$

$$m = 2$$

19.
$$8(2x + 5) = 10(x + 4)$$
$$16x + 40 = 10x + 40$$
$$16x + 40 - 10x = 10x + 40 - 10x$$
$$6x + 40 = 40$$
$$6x + 40 - 40 = 40 - 40$$
$$6x = 0$$

$$\frac{6x}{6} = \frac{0}{6}$$

$$x = 0$$

20. The product of –18 and 3 yields –54.
$$\downarrow$$

$$-18 \cdot 3 \qquad\qquad = \qquad -54$$

$$-18 \cdot 3 = -54$$

21. Twice the difference of 12 and –5 amounts to 34.

$$2 \cdot [12 - (-5)] = 34$$

$$2[12 - (-5)] = 34$$

22. Let x = unknown number

In words:	four times a number	and	sum of six times the number	is	thirty
Translate:	$4x$	$+$	$6x$	$=$	30

$$4x + 6x = 30$$
$$10x = 30$$

$$\frac{10x}{10} = \frac{30}{10}$$

$$x = 3$$

The number is 3.

23. Let x = unknown number

In words:	Thirty-two less half of 32	is equal to	sum of some number	and	5
Translate:	$32 \quad - \quad \frac{1}{2} \cdot 32$	$=$	x	$+$	5

$$32 - \frac{1}{2}(32) = x + 5$$

$$32 - 16 = x + 5$$
$$16 = x + 5$$
$$16 - 5 = x + 5 - 5$$
$$11 = x$$

The number is 11.

24.   Let x = number of Jamal's free throws
then $3x$ = number of Gary's free throws

In words: total number   is   twenty-four

Translate: $x + 3x$ =   24

$$x + 3x = 24$$

$$4x = 24$$

$$\frac{4x}{4} = \frac{24}{4}$$

$$x = 6$$

Gary made $3x = 3(6) = 18$ free throws.

25. Let x = number of women
then $x + 18$ = number of men

In words: total number   is   90

Translate: $x + x + 18$ =   90

$$x + x + 18 = 90$$
$$2x + 18 = 90$$
$$2x + 18 - 18 = 90 - 18$$
$$2x = 72$$

$$\frac{2x}{2} = \frac{72}{2}$$

$$x = 36$$

There are 36 women in the league.

4.1 SOLUTIONS TO EXERCISES

1. $\dfrac{3 \text{ shaded}}{8 \text{ total}} = \dfrac{3}{8}$

2. $\dfrac{2 \text{ shaded}}{9 \text{ total}} = \dfrac{2}{9}$

3. $\dfrac{4 \text{ shaded}}{3 \text{ total per figure}} = \dfrac{4}{3}$

4. $\dfrac{7 \text{ shaded}}{4 \text{ total per figure}} = \dfrac{7}{4}$

5.

6.

7.

8. $\dfrac{7}{9} = \dfrac{7 \cdot 5}{9 \cdot 5} = \dfrac{35}{45}$

9. $\dfrac{4}{5} = \dfrac{4 \cdot 12}{5 \cdot 12} = \dfrac{48}{60}$

10. $\dfrac{6a}{11} = \dfrac{6a \cdot 3}{11 \cdot 3} = \dfrac{18a}{33}$

11. $\dfrac{8y}{13} = \dfrac{8y \cdot 2}{13 \cdot 2} = \dfrac{16y}{26}$

12. $\dfrac{4}{9b} = \dfrac{4 \cdot 4}{9b \cdot 4} = \dfrac{16}{36b}$

13. $6 = \dfrac{6}{1} = \dfrac{6 \cdot 9}{1 \cdot 9} = \dfrac{54}{9}$

14. $2 = \dfrac{2}{1} = \dfrac{2 \cdot 15}{1 \cdot 15} = \dfrac{30}{15}$

15. $\dfrac{11}{11} = 1$

16. $\dfrac{-9}{-9} = 1$

17. $\dfrac{27}{9} = 3$

18. $\dfrac{-36}{12} = -3$

19. $\dfrac{12}{-12} = -1$

20. $\dfrac{30}{1} = 30$

4.2 SOLUTIONS TO EXERCISES

1. $84 = 2 \cdot 42$
   $= 2 \cdot 2 \cdot 21$
   $= 2 \cdot 2 \cdot 3 \cdot 7$
   or $2^2 \cdot 3 \cdot 7$

2. $180 = 2 \cdot 90$
   $= 2 \cdot 2 \cdot 45$
   $= 2 \cdot 2 \cdot 3 \cdot 15$
   $= 2 \cdot 2 \cdot 3 \cdot 3 \cdot 5$
   or $2^2 \cdot 3^2 \cdot 5$

3. $120 = 2 \cdot 60$
   $= 2 \cdot 2 \cdot 30$
   $= 2 \cdot 2 \cdot 2 \cdot 15$
   $= 2 \cdot 2 \cdot 2 \cdot 3 \cdot 5$
   or $2^3 \cdot 3 \cdot 5$

4. $504 = 2 \cdot 252$
   $= 2 \cdot 2 \cdot 126$
   $= 2 \cdot 2 \cdot 2 \cdot 63$
   $= 2 \cdot 2 \cdot 2 \cdot 7 \cdot 9$
   $= 2 \cdot 2 \cdot 2 \cdot 7 \cdot 3 \cdot 3$
   or $2^3 \cdot 3^2 \cdot 7$

5. $\dfrac{14}{21} = \dfrac{2 \cdot 7}{3 \cdot 7}$

       $= \dfrac{2}{3}$

6. $\dfrac{80}{90} = \dfrac{2 \cdot 2 \cdot 2 \cdot 2 \cdot 5}{2 \cdot 5 \cdot 3 \cdot 3}$

       $= \dfrac{8}{9}$

7. $\dfrac{15}{35} = \dfrac{3 \cdot 5}{5 \cdot 7}$

       $= \dfrac{3}{7}$

8. $\dfrac{28}{6} = \dfrac{2 \cdot 2 \cdot 7}{2 \cdot 3}$

       $= \dfrac{14}{3}$

9. $\dfrac{70}{42} = \dfrac{2 \cdot 5 \cdot 7}{2 \cdot 3 \cdot 7}$

       $= \dfrac{5}{3}$

10. $\dfrac{121}{33} = \dfrac{11 \cdot 11}{3 \cdot 11}$

       $= \dfrac{11}{3}$

11. $\dfrac{18b}{45b} = \dfrac{2 \cdot 3 \cdot 3 \cdot b}{3 \cdot 3 \cdot 5 \cdot b}$

       $= \dfrac{2}{5}$

12. $\dfrac{12xy}{52x} = \dfrac{2 \cdot 2 \cdot 3 \cdot x \cdot y}{2 \cdot 2 \cdot 13 \cdot x}$

       $= \dfrac{3y}{13}$

13. $\dfrac{19x^2}{38x} = \dfrac{19 \cdot x \cdot x}{2 \cdot 19 \cdot x}$

       $= \dfrac{x}{2}$

14. $\dfrac{51x}{85xy} = \dfrac{3 \cdot 17 \cdot x}{5 \cdot 17 \cdot x \cdot y}$

       $= \dfrac{3}{5y}$

15. $\dfrac{90m^2}{350mn} = \dfrac{2 \cdot 3 \cdot 3 \cdot 5 \cdot m \cdot m}{2 \cdot 5 \cdot 5 \cdot 7 \cdot m \cdot n}$

       $= \dfrac{9m}{35n}$

16. $\dfrac{8xyz}{54xz} = \dfrac{2 \cdot 2 \cdot 2 \cdot x \cdot y \cdot z}{2 \cdot 3 \cdot 3 \cdot 3 \cdot x \cdot z}$

       $= \dfrac{4y}{27}$

17. $\dfrac{144ab}{84b^2} = \dfrac{2 \cdot 2 \cdot 2 \cdot 2 \cdot 3 \cdot 3 \cdot a \cdot b}{2 \cdot 2 \cdot 3 \cdot 7 \cdot b \cdot b}$

       $= \dfrac{12a}{7b}$

18. $\dfrac{504x}{45x^2} = \dfrac{2 \cdot 2 \cdot 2 \cdot 3 \cdot 3 \cdot 7 \cdot x}{3 \cdot 3 \cdot 5 \cdot x \cdot x}$

       $= \dfrac{56}{5x}$

19. $\dfrac{220 \text{ yards}}{1760 \text{ yards}} = \dfrac{2 \cdot 2 \cdot 5 \cdot 11}{2 \cdot 2 \cdot 2 \cdot 2 \cdot 2 \cdot 5 \cdot 11}$

           $= \dfrac{1}{8}$

$\dfrac{1}{8}$ of a mile is represented.

20. $\dfrac{6 \text{ hours}}{40 \text{ hours}} = \dfrac{2 \cdot 3}{2 \cdot 2 \cdot 2 \cdot 5}$

$\qquad\qquad = \dfrac{3}{20}$

$\dfrac{3}{20}$ of a work week is represented.

4.3 SOLUTIONS TO EXERCISES

1. $\dfrac{3}{4} \cdot \dfrac{10}{33} = \dfrac{3 \cdot 10}{4 \cdot 33}$

$\qquad\quad = \dfrac{3 \cdot 2 \cdot 5}{2 \cdot 2 \cdot 3 \cdot 11}$

$\qquad\quad = \dfrac{5}{22}$

2. $-\dfrac{5}{6} \cdot \dfrac{4}{7} = -\dfrac{5 \cdot 4}{6 \cdot 7}$

$\qquad\quad = -\dfrac{5 \cdot 2 \cdot 2}{2 \cdot 3 \cdot 7}$

$\qquad\quad = -\dfrac{10}{21}$

3. $\dfrac{8y}{11} \cdot \dfrac{22}{18} = \dfrac{8y \cdot 22}{11 \cdot 18}$

$\qquad\quad = \dfrac{2 \cdot 2 \cdot 2 \cdot y \cdot 2 \cdot 11}{11 \cdot 2 \cdot 3 \cdot 3}$

$\qquad\quad = \dfrac{8y}{9}$

4. $9 \cdot \dfrac{1}{2} = \dfrac{9}{1} \cdot \dfrac{1}{2}$

$\qquad\quad = \dfrac{9 \cdot 1}{1 \cdot 2}$

$\qquad\quad = \dfrac{9}{2}$

5. $-\dfrac{5}{6} \cdot 12 = -\dfrac{5}{6} \cdot \dfrac{12}{1}$

$\qquad\qquad = -\dfrac{5 \cdot 12}{6 \cdot 1}$

$\qquad\qquad = -\dfrac{5 \cdot 2 \cdot 6}{6 \cdot 1}$

$\qquad\qquad = -10$

6. $\dfrac{5}{3} \div \dfrac{1}{6} = \dfrac{5}{3} \cdot \dfrac{6}{1}$

$\qquad\qquad = \dfrac{5 \cdot 2 \cdot 3}{3 \cdot 1}$

$\qquad\qquad = 10$

7. $\dfrac{20}{33} \div -\dfrac{5}{11} = \dfrac{20}{33} \cdot -\dfrac{11}{5}$

$\qquad\qquad = -\dfrac{20 \cdot 11}{33 \cdot 5}$

$\qquad\qquad = -\dfrac{2 \cdot 2 \cdot 5 \cdot 11}{3 \cdot 11 \cdot 5}$

$\qquad\qquad = -\dfrac{4}{3}$

8. $\dfrac{12y}{13} \div \dfrac{3}{26} = \dfrac{12y}{13} \cdot \dfrac{26}{3}$

$\qquad\qquad = \dfrac{12y \cdot 26}{13 \cdot 3}$

$\qquad\qquad = \dfrac{2 \cdot 2 \cdot 3 \cdot y \cdot 2 \cdot 13}{13 \cdot 3}$

$\qquad\qquad = 8y$

9. $-\dfrac{7}{8} \div 14 = -\dfrac{7}{8} \div \dfrac{14}{1}$

$\qquad\qquad = -\dfrac{7}{8} \cdot \dfrac{1}{14}$

$\qquad\qquad = -\dfrac{7 \cdot 1}{8 \cdot 14}$

$\qquad\qquad = -\dfrac{7 \cdot 1}{8 \cdot 7 \cdot 2}$

$\qquad\qquad = -\dfrac{1}{16}$

10. $\left(-\dfrac{7}{8}\right)^2 = \left(-\dfrac{7}{8}\right)\left(-\dfrac{7}{8}\right)$

$= \dfrac{7 \cdot 7}{8 \cdot 8}$

$= \dfrac{49}{64}$

11. $\left(\dfrac{2}{3}\right)^5 = \dfrac{2}{3} \cdot \dfrac{2}{3} \cdot \dfrac{2}{3} \cdot \dfrac{2}{3} \cdot \dfrac{2}{3}$

$= \dfrac{2 \cdot 2 \cdot 2 \cdot 2 \cdot 2}{3 \cdot 3 \cdot 3 \cdot 3 \cdot 3}$

$= \dfrac{32}{243}$

12. $\dfrac{1}{4} \cdot \dfrac{2}{5} \div \dfrac{3}{10} = \dfrac{1 \cdot 2}{2 \cdot 2 \cdot 5} \div \dfrac{3}{10}$

$= \dfrac{1}{10} \div \dfrac{3}{10}$

$= \dfrac{1}{10} \cdot \dfrac{10}{3}$

$= \dfrac{1 \cdot 10}{10 \cdot 3}$

$= \dfrac{1}{3}$

13. $\dfrac{4}{9} \div \dfrac{2}{27} \cdot -\dfrac{2}{5} = \dfrac{4}{9} \cdot \dfrac{27}{2} \cdot -\dfrac{2}{5}$

$= -\dfrac{2 \cdot 2 \cdot 3 \cdot 3 \cdot 3 \cdot 2}{3 \cdot 3 \cdot 2 \cdot 5}$

$= -\dfrac{12}{5}$

14. $12x \div \dfrac{24x}{5} = \dfrac{12x}{1} \cdot \dfrac{5}{24x}$

$= -\dfrac{12 \cdot x \cdot 5}{1 \cdot 2 \cdot 12 \cdot x}$

$= \dfrac{5}{2}$

15. $\left(25 \div \dfrac{5}{4}\right) \cdot \dfrac{3}{16} = \left(\dfrac{25}{1} \cdot \dfrac{4}{5}\right) \cdot \dfrac{3}{16}$

$= \left(\dfrac{5 \cdot 5 \cdot 4}{1 \cdot 5}\right) \cdot \dfrac{3}{16}$

$= \dfrac{5 \cdot 4}{1} \cdot \dfrac{3}{4 \cdot 4}$

$= \dfrac{15}{4}$

16. (a) $xy = \left(\dfrac{3}{7}\right)\left(\dfrac{7}{9}\right)$

$= \dfrac{3 \cdot 7}{7 \cdot 3 \cdot 3}$

$= \dfrac{1}{3}$

(b) $x \div y = \dfrac{3}{7} \div \dfrac{7}{9}$

$= \dfrac{3}{7} \cdot \dfrac{9}{7}$

$= \dfrac{3 \cdot 9}{7 \cdot 7}$

$= \dfrac{27}{49}$

17. (a) $xy = \left(-\dfrac{8}{15}\right)\left(\dfrac{3}{4}\right)$

$= -\dfrac{8 \cdot 3}{15 \cdot 4}$

$= -\dfrac{4 \cdot 2 \cdot 3}{3 \cdot 5 \cdot 4}$

$= -\dfrac{2}{5}$

(b) $x \div y = -\dfrac{8}{15} \div \dfrac{3}{4}$

$= -\dfrac{8}{15} \cdot \dfrac{4}{3}$

$= -\dfrac{8 \cdot 4}{15 \cdot 3}$

$= -\dfrac{32}{45}$

18. $$2x = -\frac{5}{8}$$

$$2\left(-\frac{5}{16}\right) = -\frac{5}{8}$$

$$\frac{2}{1}\left(-\frac{5}{16}\right) = -\frac{5}{8}$$

$$-\frac{2 \cdot 5}{1 \cdot 16} = -\frac{5}{8}$$

$$-\frac{2 \cdot 5}{1 \cdot 2 \cdot 8} = -\frac{5}{8}$$

$$-\frac{5}{8} = -\frac{5}{8}$$

True

$-\frac{5}{16}$ is a solution.

19. $$-\frac{3}{4}y = \frac{1}{12}$$

$$-\frac{3}{4}\left(\frac{1}{9}\right) = \frac{1}{12}$$

$$-\frac{3 \cdot 1}{4 \cdot 9} = \frac{1}{12}$$

$$-\frac{3 \cdot 1}{4 \cdot 3 \cdot 3} = \frac{1}{12}$$

$$-\frac{1}{12} = \frac{1}{12}$$

False

$\frac{1}{9}$ is not a solution.

20. fraction of orchard · total number = number of
     that are apple trees   of trees   apple trees

$\frac{5}{8}$ ·   64   =   number of
                                                       apple trees

$$\frac{5}{8} \cdot 64 = \frac{5}{8} \cdot \frac{64}{1}$$

$$= \frac{5 \cdot 8 \cdot 8}{8 \cdot 1}$$

$$= 40$$

There are 40 apple trees.

4.4 SOLUTIONS TO EXERCISES

1. $$\frac{1}{8} + \frac{5}{8} = \frac{1 + 5}{8}$$

$$= \frac{6}{8}$$

$$= \frac{2 \cdot 3}{2 \cdot 2 \cdot 2}$$

$$= \frac{3}{4}$$

2. $$\frac{5}{12} + \frac{7}{12} = \frac{5 + 7}{12}$$

$$= \frac{12}{12}$$

$$= 1$$

3. $$-\frac{1}{3} + \frac{1}{3} = \frac{-1 + 1}{3}$$

$$= \frac{0}{3}$$

$$= 0$$

4. $$-\frac{2}{x} + \frac{6}{x} = \frac{-2 + 6}{x}$$

$$= \frac{4}{x}$$

5. $\dfrac{7}{10y} + \dfrac{1}{10y} = \dfrac{7 + 1}{10y}$

$\qquad\qquad = \dfrac{8}{10y}$

$\qquad\qquad = \dfrac{2 \cdot 2 \cdot 2}{2 \cdot 5 \cdot y}$

$\qquad\qquad = \dfrac{4}{5y}$

6. $\dfrac{3}{17} + \dfrac{2}{17} + \dfrac{5}{17} = \dfrac{3 + 2 + 5}{17}$

$\qquad\qquad\qquad = \dfrac{10}{17}$

7. $\dfrac{9}{13} - \dfrac{5}{13} = \dfrac{9 - 5}{13}$

$\qquad\qquad = \dfrac{4}{13}$

8. $\dfrac{2}{y} - \dfrac{8}{y} = \dfrac{2 - 8}{y}$

$\qquad\quad = \dfrac{-6}{y}$

$\qquad\quad$ or $-\dfrac{6}{y}$

9. $\dfrac{5}{22} - \dfrac{3}{22} = \dfrac{5 - 3}{22}$

$\qquad\qquad = \dfrac{2}{22}$

$\qquad\qquad = \dfrac{2}{2 \cdot 11}$

$\qquad\qquad = \dfrac{1}{11}$

10. $\dfrac{8}{15a} + \dfrac{13}{15a} = \dfrac{8 + 13}{15a}$

$\qquad\qquad = \dfrac{21}{15a}$

$\qquad\qquad = \dfrac{3 \cdot 7}{3 \cdot 5 \cdot a}$

$\qquad\qquad = \dfrac{7}{5a}$

11. $-\dfrac{5}{14} + \dfrac{3}{14} = \dfrac{-5 + 3}{14}$

$\qquad\qquad = \dfrac{-2}{14}$

$\qquad\qquad = -\dfrac{2}{2 \cdot 7}$

$\qquad\qquad = -\dfrac{1}{7}$

12. $\dfrac{11x}{16} - \dfrac{17x}{16} = \dfrac{11x - 17x}{16}$

$\qquad\qquad = \dfrac{-6x}{16}$

$\qquad\qquad = -\dfrac{2 \cdot 3 \cdot x}{2 \cdot 2 \cdot 2 \cdot 2}$

$\qquad\qquad = -\dfrac{3x}{8}$

13. $\quad x + \dfrac{5}{8} = \dfrac{1}{2}$

$\quad -\dfrac{1}{8} + \dfrac{5}{8} = \dfrac{1}{2}$

$\quad \dfrac{-1 + 5}{8} = \dfrac{1}{2}$

$\quad\quad \dfrac{4}{8} = \dfrac{1}{2}$

$\quad\quad \dfrac{1}{2} = \dfrac{1}{2}$

$\qquad$ True

$-\dfrac{1}{8}$ is a solution.

14. $-\dfrac{3}{4} + y = -\dfrac{1}{8}$

$-\dfrac{3}{4} + \dfrac{1}{4} = -\dfrac{1}{8}$

$\dfrac{-3 + 1}{4} = -\dfrac{1}{8}$

$\dfrac{-2}{4} = -\dfrac{1}{8}$

$-\dfrac{1}{2} = -\dfrac{1}{8}$

False

$\dfrac{1}{4}$ is not a solution.

15. $6 = 2 \cdot 3$
$27 = 3 \cdot 3 \cdot 3$

LCD $= 2 \cdot 3 \cdot 3 \cdot 3 = 54$

16. $8 = 2 \cdot 2 \cdot 2$
$36 = 2 \cdot 2 \cdot 3 \cdot 3$

LCD $= 2 \cdot 2 \cdot 2 \cdot 3 \cdot 3 = 72$

17. $x + \dfrac{1}{10} = -\dfrac{7}{10}$

$x + \dfrac{1}{10} + \left(-\dfrac{1}{10}\right) = -\dfrac{7}{10} + \left(-\dfrac{1}{10}\right)$

$x = \dfrac{-7 + (-1)}{10}$

$x = \dfrac{-8}{10}$

$x = -\dfrac{4}{5}$

18. $4x + \dfrac{1}{13} - 3x = \dfrac{3}{13} - \dfrac{9}{13}$

$x + \dfrac{1}{13} = \dfrac{3 - 9}{13}$

$x + \dfrac{1}{13} = \dfrac{-6}{13}$

$x + \dfrac{1}{13} + \left(-\dfrac{1}{13}\right) = \dfrac{-6}{13} + \left(-\dfrac{1}{13}\right)$

$x = \dfrac{-6 + (-1)}{13}$

$x = \dfrac{-7}{13}$ or $-\dfrac{7}{13}$

19.

1st amount of flour	+	2nd amount of flour	=	total amount of flour
$\dfrac{4}{3}$	+	$\dfrac{2}{3}$	=	$\dfrac{4 + 2}{3}$

$= \dfrac{6}{3}$

$= 2$

It uses 2 cups of flour.

20.

amount of milk in pitcher	+	amount added	−	amount poured on cereal	=	amount left in pitcher
$\dfrac{4}{8}$	+	$\dfrac{1}{8}$	−	$\dfrac{2}{8}$	=	$\dfrac{4 + 1 - 2}{8}$

$= \dfrac{3}{8}$

$\dfrac{3}{8}$ cup of milk was left in the pitcher.

4.5 SOLUTIONS TO EXERCISES

1. $\dfrac{3}{8} + \dfrac{1}{16} = \dfrac{3 \cdot 2}{8 \cdot 2} + \dfrac{1}{16}$

$= \dfrac{6}{16} + \dfrac{1}{16}$

$= \dfrac{7}{16}$

2. $\dfrac{4}{5} - \dfrac{7}{10} = \dfrac{4 \cdot 2}{5 \cdot 2} - \dfrac{7}{10}$

$= \dfrac{8}{10} - \dfrac{7}{10}$

$= \dfrac{1}{10}$

3. $-\dfrac{2}{9} + \dfrac{5}{3} = -\dfrac{2}{9} + \dfrac{5 \cdot 3}{3 \cdot 3}$

$= -\dfrac{2}{9} + \dfrac{15}{9}$

$= \dfrac{13}{9}$

4. $\dfrac{7x}{13} - \dfrac{5}{26} = \dfrac{7x \cdot 2}{13 \cdot 2} - \dfrac{5}{26}$

$= \dfrac{14x}{26} - \dfrac{5}{26}$

$= \dfrac{14x - 5}{26}$

5. $\dfrac{5y}{6} - \dfrac{5}{12} = \dfrac{5y \cdot 2}{6 \cdot 2} - \dfrac{5}{12}$

$= \dfrac{10y}{12} - \dfrac{5}{12}$

$= \dfrac{10y - 5}{12}$

6. $9x - \dfrac{3}{10} = \dfrac{9x}{1} - \dfrac{3}{10}$

$= \dfrac{9x \cdot 10}{1 \cdot 10} - \dfrac{3}{10}$

$= \dfrac{90x}{10} - \dfrac{3}{10}$

$= \dfrac{90x - 3}{10}$

7. $\dfrac{9}{20} + \dfrac{5}{20} + \dfrac{3}{20} = \dfrac{9 + 5 + 3}{20}$

$= \dfrac{17}{20}$

8. $\dfrac{8}{23} + \dfrac{31}{23} - \dfrac{16}{23} = \dfrac{8 + 31 - 16}{23}$

$= \dfrac{23}{23}$

$= 1$

9. $-\dfrac{4}{5} + \dfrac{9}{10} - \dfrac{1}{15} = \dfrac{-4 \cdot 6}{5 \cdot 6} + \dfrac{9 \cdot 3}{10 \cdot 3} - \dfrac{1 \cdot 2}{15 \cdot 2}$

$= \dfrac{-24}{30} + \dfrac{27}{30} - \dfrac{2}{30}$

$= \dfrac{-24 + 27 - 2}{30}$

$= \dfrac{1}{30}$

10. $\dfrac{y}{2} + \dfrac{y}{6} + \dfrac{5y}{12} = \dfrac{y \cdot 6}{2 \cdot 6} + \dfrac{y \cdot 2}{6 \cdot 2} + \dfrac{5y}{12}$

$= \dfrac{6y}{12} + \dfrac{2y}{12} + \dfrac{5y}{12}$

$= \dfrac{6y + 2y + 5y}{12}$

$= \dfrac{13y}{12}$

11. $\dfrac{5}{7} + \dfrac{6}{11x} = \dfrac{5 \cdot 11x}{7 \cdot 11x} + \dfrac{6 \cdot 7}{11x \cdot 7}$

$= \dfrac{55x}{77x} + \dfrac{42}{77x}$

$= \dfrac{55x + 42}{77x}$

12. $\dfrac{4}{15x} + \dfrac{3}{4} = \dfrac{4 \cdot 4}{15x \cdot 4} + \dfrac{3 \cdot 15x}{4 \cdot 15x}$

$= \dfrac{16}{60x} + \dfrac{45x}{60x}$

$= \dfrac{16 + 45x}{60x}$

13. $\dfrac{5}{12} + \dfrac{9b}{16} - \dfrac{7}{8} = \dfrac{5 \cdot 4}{12 \cdot 4} + \dfrac{9b \cdot 3}{16 \cdot 3} - \dfrac{7 \cdot 6}{8 \cdot 6}$

$= \dfrac{20}{48} + \dfrac{27b}{48} - \dfrac{42}{48}$

$= \dfrac{20 + 27b - 42}{48}$

$= \dfrac{27b - 22}{48}$

14. $\dfrac{4x}{13} - \dfrac{3x}{26} - \dfrac{5}{2} = \dfrac{4x \cdot 2}{13 \cdot 2} - \dfrac{3x}{26} - \dfrac{5 \cdot 13}{2 \cdot 13}$

$\qquad\qquad = \dfrac{8x}{26} - \dfrac{3x}{26} - \dfrac{65}{26}$

$\qquad\qquad = \dfrac{8x - 3x - 65}{26}$

$\qquad\qquad = \dfrac{5x - 65}{26}$

15. $2x + y = 2\left(\dfrac{1}{4}\right) + \dfrac{2}{5}$

$\qquad\qquad = \dfrac{2}{1} \cdot \dfrac{1}{4} + \dfrac{2}{5}$

$\qquad\qquad = \dfrac{1}{2} + \dfrac{2}{5}$

$\qquad\qquad = \dfrac{1 \cdot 5}{2 \cdot 5} + \dfrac{2 \cdot 2}{5 \cdot 2}$

$\qquad\qquad = \dfrac{5}{10} + \dfrac{4}{10}$

$\qquad\qquad = \dfrac{9}{10}$

16. $x - y = \dfrac{1}{4} - \dfrac{2}{5}$

$\qquad\quad = \dfrac{1 \cdot 5}{4 \cdot 5} - \dfrac{2 \cdot 4}{5 \cdot 4}$

$\qquad\quad = \dfrac{5}{20} - \dfrac{8}{20}$

$\qquad\quad = \dfrac{-3}{20} \;\; \text{or} \;\; -\dfrac{3}{20}$

17. $\qquad x - \dfrac{8}{9} = \dfrac{2}{3}$

$\quad x - \dfrac{8}{9} + \dfrac{8}{9} = \dfrac{2}{3} + \dfrac{8}{9}$

$\qquad\qquad x = \dfrac{2 \cdot 3}{3 \cdot 3} + \dfrac{8}{9}$

$\qquad\qquad x = \dfrac{6}{9} + \dfrac{8}{9}$

$\qquad\qquad x = \dfrac{14}{9}$

18. $15y - \dfrac{3}{7} - 14y = \dfrac{13}{14}$

$\qquad\qquad y - \dfrac{3}{7} = \dfrac{13}{14}$

$\quad y - \dfrac{3}{7} + \dfrac{3}{7} = \dfrac{13}{14} + \dfrac{3}{7}$

$\qquad\qquad y = \dfrac{13}{14} + \dfrac{3 \cdot 2}{7 \cdot 2}$

$\qquad\qquad y = \dfrac{13}{14} + \dfrac{6}{14}$

$\qquad\qquad y = \dfrac{19}{14}$

19.

weight of printers	+	weight of copiers	+	weight of stereos	=	total weight
$\dfrac{2}{5}$	+	$\dfrac{1}{2}$	+	$\dfrac{1}{3}$	=	$\dfrac{2 \cdot 6}{5 \cdot 6} + \dfrac{1 \cdot 15}{2 \cdot 15} + \dfrac{1 \cdot 10}{3 \cdot 10}$

$\qquad\qquad\qquad\qquad\qquad\qquad = \dfrac{12}{30} + \dfrac{15}{30} + \dfrac{10}{30}$

$\qquad\qquad\qquad\qquad\qquad\qquad = \dfrac{37}{30}$

The total weight of the load is $\dfrac{37}{30}$ tons.

20.

1st length	−	2nd length	=	difference in lengths
$\dfrac{5}{6}$	−	$\dfrac{3}{4}$	=	$\dfrac{5 \cdot 2}{6 \cdot 2} - \dfrac{3 \cdot 3}{4 \cdot 3}$

$\qquad\qquad\qquad\qquad\qquad = \dfrac{10}{12} - \dfrac{9}{12}$

$\qquad\qquad\qquad\qquad\qquad = \dfrac{1}{12}$

The difference is $\dfrac{1}{12}$ of a foot.

4.6 SOLUTIONS TO EXERCISES

1. $\dfrac{\frac{8}{27}}{\frac{1}{9}} = \dfrac{8}{27} \div \dfrac{1}{9}$

    $= \dfrac{8}{27} \cdot \dfrac{9}{1}$

    $= \dfrac{8 \cdot 9}{9 \cdot 3 \cdot 1}$

    $= \dfrac{8}{3}$

2. $\dfrac{\frac{2x}{13}}{\frac{4}{3}} = \dfrac{2x}{13} \div \dfrac{4}{3}$

    $= \dfrac{2x}{13} \cdot \dfrac{3}{4}$

    $= \dfrac{2 \cdot x \cdot 3}{13 \cdot 2 \cdot 2}$

    $= \dfrac{3x}{26}$

3. $\dfrac{\frac{3}{8} + \frac{1}{4}}{\frac{2}{5} + \frac{7}{10}} = \dfrac{40\left(\frac{3}{8} + \frac{1}{4}\right)}{40\left(\frac{2}{5} + \frac{7}{10}\right)}$

    $= \dfrac{40\left(\frac{3}{8}\right) + 40\left(\frac{1}{4}\right)}{40\left(\frac{2}{5}\right) + 40\left(\frac{7}{10}\right)}$

    $= \dfrac{15 + 10}{16 + 28}$

    $= \dfrac{25}{44}$

4. $\dfrac{\frac{7x}{5}}{6 - \frac{1}{10}} = \dfrac{10\left(\frac{7x}{5}\right)}{10\left(6 - \frac{1}{10}\right)}$

    $= \dfrac{10\left(\frac{7x}{5}\right)}{10(6) - 10\left(\frac{1}{10}\right)}$

    $= \dfrac{14x}{60 - 1}$

    $= \dfrac{14x}{59}$

5. $\dfrac{\frac{6}{11} + 1}{\frac{3}{8a}} = \dfrac{88a\left(\frac{6}{11} + 1\right)}{88a\left(\frac{3}{8a}\right)}$

    $= \dfrac{88a\left(\frac{6}{11}\right) + 88a(1)}{88a\left(\frac{3}{8a}\right)}$

    $= \dfrac{48a + 88a}{33}$

6. $\dfrac{9 - \frac{1}{3}}{7 + \frac{3}{5}} = \dfrac{15\left(9 - \frac{1}{3}\right)}{15\left(7 + \frac{3}{5}\right)}$

    $= \dfrac{15(9) - 15\left(\frac{1}{3}\right)}{15(7) + 15\left(\frac{3}{5}\right)}$

    $= \dfrac{135 - 5}{105 + 9}$

    $= \dfrac{130}{114}$

    $= \dfrac{2 \cdot 65}{2 \cdot 57}$

    $= \dfrac{65}{57}$

7. $\left(-\dfrac{3}{8} - \dfrac{9}{8}\right) \div \dfrac{5}{16} = -\dfrac{12}{8} \div \dfrac{5}{16}$

$\qquad\qquad = -\dfrac{3}{2} \cdot \dfrac{16}{5}$

$\qquad\qquad = -\dfrac{3 \cdot 2 \cdot 8}{2 \cdot 5}$

$\qquad\qquad = -\dfrac{24}{5}$

8. $5^2 - \left(\dfrac{4}{3}\right)^2 = 25 - \dfrac{16}{9}$

$\qquad\qquad = \dfrac{225}{9} - \dfrac{16}{9}$

$\qquad\qquad = \dfrac{209}{9}$

9. $\left(3 - \dfrac{5}{2}\right)^2 = \left(\dfrac{6}{2} - \dfrac{5}{2}\right)^2$

$\qquad\qquad = \left(\dfrac{1}{2}\right)^2$

$\qquad\qquad = \dfrac{1}{4}$

10. $\left(\dfrac{4}{5} - 1\right)\left(\dfrac{3}{4} + \dfrac{7}{8}\right) = \left(\dfrac{4}{5} - \dfrac{5}{5}\right)\left(\dfrac{6}{8} + \dfrac{7}{8}\right)$

$\qquad\qquad = \left(-\dfrac{1}{5}\right)\left(\dfrac{13}{8}\right)$

$\qquad\qquad = -\dfrac{13}{40}$

11. $\left(\dfrac{4}{5} \cdot \dfrac{5}{9}\right) - \left(\dfrac{2}{3} \div \dfrac{3}{7}\right) = \left(\dfrac{4 \cdot 5}{5 \cdot 9}\right) - \left(\dfrac{2}{3} \cdot \dfrac{7}{3}\right)$

$\qquad\qquad = \dfrac{4}{9} - \dfrac{14}{9}$

$\qquad\qquad = -\dfrac{10}{9}$

12. $\left(-\dfrac{8}{7} + \dfrac{1}{7}\right)^5 = \left(-\dfrac{7}{7}\right)^5$

$\qquad\qquad = (-1)^5$

$\qquad\qquad = -1$

13. $\dfrac{\dfrac{x}{4} + 5}{2 + \dfrac{2}{3}} = \dfrac{12\left(\dfrac{x}{4} + 5\right)}{12\left(2 + \dfrac{2}{3}\right)}$

$\qquad\qquad = \dfrac{12\left(\dfrac{x}{4}\right) + 12(5)}{12(2) + 12\left(\dfrac{2}{3}\right)}$

$\qquad\qquad = \dfrac{3x + 60}{24 + 8}$

$\qquad\qquad = \dfrac{3x + 60}{32}$

14. $\dfrac{7 - \dfrac{x}{6}}{10 + \dfrac{5}{12}} = \dfrac{12\left(7 - \dfrac{x}{6}\right)}{12\left(10 + \dfrac{5}{12}\right)}$

$\qquad\qquad = \dfrac{12(7) - 12\left(\dfrac{x}{6}\right)}{12(10) + 12\left(\dfrac{5}{12}\right)}$

$\qquad\qquad = \dfrac{84 - 2x}{120 + 5}$

$\qquad\qquad = \dfrac{84 - 2x}{125}$

15. $\dfrac{1 + x}{y} = \dfrac{1 + \dfrac{2}{5}}{-\dfrac{5}{6}}$

$\qquad\qquad = \dfrac{30\left(1 + \dfrac{2}{5}\right)}{30\left(-\dfrac{5}{6}\right)}$

$$= \frac{30(1) + 30\left(\frac{2}{5}\right)}{30\left(-\frac{5}{6}\right)}$$

$$= \frac{30 + 12}{-25}$$

$$= -\frac{42}{25}$$

16. $10x - y = 10\left(\frac{2}{5}\right) - \left(-\frac{5}{6}\right)$

$$= 4 + \frac{5}{6}$$

$$= \frac{24}{6} + \frac{5}{6}$$

$$= \frac{29}{6}$$

17. $x^2 + y = \left(\frac{2}{5}\right)^2 + \left(-\frac{5}{6}\right)$

$$= \frac{4}{25} + \left(-\frac{5}{6}\right)$$

$$= \frac{24}{150} + \left(-\frac{125}{150}\right)$$

$$= -\frac{101}{150}$$

18. $xy = \left(\frac{2}{5}\right)\left(-\frac{5}{6}\right)$

$$= -\frac{2 \cdot 5}{5 \cdot 2 \cdot 3}$$

$$= -\frac{1}{3}$$

19. $\dfrac{\dfrac{4}{9} + \dfrac{7}{18}}{2} = \dfrac{18\left(\dfrac{4}{9} + \dfrac{7}{18}\right)}{18(2)}$

$$= \frac{18\left(\frac{4}{9}\right) + 18\left(\frac{7}{18}\right)}{18(2)}$$

$$= \frac{8 + 7}{36}$$

$$= \frac{15}{36}$$

$$= \frac{5}{12}$$

20. $\dfrac{\dfrac{5}{6} + \dfrac{7}{18}}{2} = \dfrac{18\left(\dfrac{5}{6} + \dfrac{7}{18}\right)}{18(2)}$

$$= \frac{18\left(\frac{5}{6}\right) + 18\left(\frac{7}{18}\right)}{18(2)}$$

$$= \frac{15 + 7}{36}$$

$$= \frac{22}{36}$$

$$= \frac{11}{18}$$

4.7 SOLUTIONS TO EXERCISES

1. $8x = 3$

$$\frac{8x}{8} = \frac{3}{8}$$

$$x = \frac{3}{8}$$

2. $-6y = 5$

$$\frac{-6y}{-6} = \frac{5}{-6}$$

$$y = -\frac{5}{6}$$

3. $\quad \frac{2}{5}a = 10$

$$\frac{5}{2} \cdot \frac{2}{5}a = \frac{5}{2} \cdot \frac{10}{1}$$

$$a = 25$$

4. $-\dfrac{5}{6}m = -\dfrac{7}{12}$

$$-\dfrac{6}{5}\left(-\dfrac{5}{6}m\right) = -\dfrac{6}{5}\left(-\dfrac{7}{12}\right)$$

$$m = \dfrac{7}{10}$$

5. $\dfrac{x}{4} + 3 = \dfrac{7}{2}$

$$4\left(\dfrac{x}{4} + 3\right) = 4\left(\dfrac{7}{2}\right)$$

$$4\left(\dfrac{x}{4}\right) + 4(3) = 4\left(\dfrac{7}{2}\right)$$

$$x + 12 = 14$$
$$x + 12 - 12 = 14 - 12$$
$$x = 2$$

6. $\dfrac{y}{8} - \dfrac{8}{16} = 1$

$$16\left(\dfrac{y}{8} - \dfrac{8}{16}\right) = 16(1)$$

$$16\left(\dfrac{y}{8}\right) - 16\left(\dfrac{8}{16}\right) = 16(1)$$

$$2y - 8 = 16$$
$$2y - 8 + 8 = 16 + 8$$
$$2y = 24$$

$$\dfrac{2y}{2} = \dfrac{24}{2}$$
$$y = 12$$

7. $\dfrac{b}{7} - b = -18$

$$7\left(\dfrac{b}{7} - b\right) = 7(-18)$$

$$7\left(\dfrac{b}{7}\right) - 7(b) = 7(-18)$$
$$b - 7b = -126$$
$$-6b = -126$$

$$\dfrac{-6b}{-6} = \dfrac{-126}{-6}$$
$$b = 21$$

8. $\dfrac{3}{10} - \dfrac{1}{3} = \dfrac{x}{20}$

$$60\left(\dfrac{3}{10} - \dfrac{1}{3}\right) = 60\left(\dfrac{x}{20}\right)$$

$$60\left(\dfrac{3}{10}\right) - 60\left(\dfrac{1}{3}\right) = 60\left(\dfrac{x}{20}\right)$$

$$18 - 20 = 3x$$
$$-2 = 3x$$

$$\dfrac{-2}{3} = \dfrac{3x}{3}$$

$$-\dfrac{2}{3} = x$$

9. $\dfrac{x}{8} - \dfrac{5}{7} = \dfrac{x \cdot 7}{8 \cdot 7} - \dfrac{5 \cdot 8}{7 \cdot 8}$

$$= \dfrac{7x}{56} - \dfrac{40}{56}$$

$$= \dfrac{7x - 40}{56}$$

10. $-\dfrac{4}{11} + \dfrac{y}{6} = -\dfrac{4 \cdot 6}{11 \cdot 6} + \dfrac{y \cdot 11}{6 \cdot 11}$

$$= \dfrac{-24}{66} + \dfrac{11y}{66}$$

$$= \dfrac{-24 + 11y}{66}$$

11. $\dfrac{3m}{4} + 9 = \dfrac{3m}{4} + \dfrac{9 \cdot 4}{1 \cdot 4}$

$$= \dfrac{3m}{4} + \dfrac{36}{4}$$

$$= \dfrac{3m + 36}{4}$$

12. $\dfrac{8a}{5} - \dfrac{7a}{10} = \dfrac{8a \cdot 2}{5 \cdot 2} - \dfrac{7a}{10}$

$$= \dfrac{16a}{10} - \dfrac{7a}{10}$$

$$= \dfrac{9a}{10}$$

13. $\dfrac{5}{9}n = \dfrac{7}{18}$

$\dfrac{9}{5} \cdot \dfrac{5}{9}n = \dfrac{9}{5} \cdot \dfrac{7}{18}$

$n = \dfrac{7}{10}$

14. $\dfrac{5}{4} + \dfrac{4}{x} = \dfrac{1}{8}$

$8x\left(\dfrac{5}{4} + \dfrac{4}{x}\right) = 8x\left(\dfrac{1}{8}\right)$

$8x\left(\dfrac{5}{4}\right) + 8x\left(\dfrac{4}{x}\right) = 8x\left(\dfrac{1}{8}\right)$

$10x + 32 = x$
$10x + 32 - 10x = x - 10x$
$32 = -9x$

$\dfrac{32}{-9} = \dfrac{-9x}{-9}$

$-\dfrac{32}{9} = x$

15. $\dfrac{11}{12} - \dfrac{5}{6} = \dfrac{11}{12} - \dfrac{5 \cdot 2}{6 \cdot 2}$

$= \dfrac{11}{12} - \dfrac{10}{12}$

$= \dfrac{1}{12}$

16. $-\dfrac{5}{9}x = \dfrac{5}{18} - \dfrac{7}{18}$

$18\left(-\dfrac{5}{9}x\right) = 18\left(\dfrac{5}{18} - \dfrac{7}{18}\right)$

$18\left(-\dfrac{5}{9}x\right) = 18\left(\dfrac{5}{18}\right) - 18\left(\dfrac{7}{18}\right)$

$-10x = 5 - 7$
$-10x = -2$

$\dfrac{-10x}{-10} = \dfrac{-2}{-10}$

$x = \dfrac{1}{5}$

17. $18 - \dfrac{14}{3} = \dfrac{18 \cdot 3}{1 \cdot 3} - \dfrac{14}{3}$

$= \dfrac{54}{3} - \dfrac{14}{3}$

$= \dfrac{40}{3}$

18. $\dfrac{y}{4} = -3 + y$

$4\left(\dfrac{y}{4}\right) = 4(-3 + y)$

$4\left(\dfrac{y}{4}\right) = 4(-3) + 4(y)$

$y = -12 + 4y$
$y - 4y = -12 + 4y - 4y$
$-3y = -12$

$\dfrac{-3y}{-3} = \dfrac{-12}{-3}$

$y = 4$

19. $\dfrac{6}{7}x = \dfrac{2}{3} - \dfrac{1}{4}$

$84\left(\dfrac{6}{7}x\right) = 84\left(\dfrac{2}{3} - \dfrac{1}{4}\right)$

$84\left(\dfrac{6}{7}x\right) = 84\left(\dfrac{2}{3}\right) - 84\left(\dfrac{1}{4}\right)$

$72x = 56 - 21$
$72x = 35$

$\dfrac{72x}{72} = \dfrac{35}{72}$

$x = \dfrac{35}{72}$

20. $\dfrac{8}{15}b = -\dfrac{4}{5} + \dfrac{1}{3}$

$15\left(\dfrac{8}{15}b\right) = 15\left(-\dfrac{4}{5} + \dfrac{1}{3}\right)$

$15\left(\dfrac{8}{15}b\right) = 15\left(-\dfrac{4}{5}\right) + 15\left(\dfrac{1}{3}\right)$

$8b = -12 + 5$

$8b = -7$

$$\frac{8b}{8} = \frac{-7}{8}$$

$$b = -\frac{7}{8}$$

4.8 SOLUTIONS TO EXERCISES

1. $9\frac{1}{4} = \dfrac{4 \cdot 9 + 1}{4}$

    $= \dfrac{37}{4}$

2. $2\frac{3}{13} = \dfrac{13 \cdot 2 + 3}{13}$

    $= \dfrac{29}{13}$

3. $12\frac{4}{5} = \dfrac{5 \cdot 12 + 4}{5}$

    $= \dfrac{64}{5}$

4.
$$\begin{array}{r} 4\frac{2}{4} \\ 4\overline{)18} \\ \underline{16} \\ 2 \end{array}$$

$$\frac{18}{4} = 4\frac{2}{4} = 4\frac{1}{2}$$

5.
$$\begin{array}{r} 19 \\ 3\overline{)57} \\ \underline{3} \\ 27 \\ \underline{27} \\ 0 \end{array}$$

$$\frac{57}{3} = 19$$

6.
$$\begin{array}{r} 7\frac{1}{10} \\ 10\overline{)71} \\ \underline{70} \\ 1 \end{array}$$

$$\frac{71}{10} = 7\frac{1}{10}$$

7. $3\frac{4}{5} \cdot \frac{1}{8} = \frac{19}{5} \cdot \frac{1}{8}$

       $= \dfrac{19}{40}$

8. $7\frac{5}{6} \cdot 3\frac{2}{3} = \frac{47}{6} \cdot \frac{11}{3}$

       $= \dfrac{517}{18}$ or $28\frac{13}{18}$

9. $\frac{4}{9} \div 6\frac{3}{4} = \frac{4}{9} \div \frac{27}{4}$

       $= \dfrac{4}{9} \cdot \dfrac{4}{27}$

       $= \dfrac{16}{243}$

10. $20\frac{3}{8} \cdot 7 = \frac{163}{8} \cdot \frac{7}{1}$

       $= \dfrac{1141}{8}$ or $142\frac{5}{8}$

11. $15\frac{2}{7} \div \frac{9}{14} = \frac{107}{7} \div \frac{9}{14}$

       $= \dfrac{107}{7} \cdot \dfrac{14}{9}$

       $= \dfrac{107 \cdot 2 \cdot 7}{7 \cdot 9}$

       $= \dfrac{214}{9}$ or $23\frac{7}{9}$

12. $12 \div 4\frac{6}{7} = \frac{12}{1} + \frac{34}{7}$

$= \frac{12}{1} \cdot \frac{7}{34}$

$= \frac{2 \cdot 6 \cdot 7}{1 \cdot 2 \cdot 17}$

$= \frac{42}{17}$ or $2\frac{8}{17}$

13. $\quad 4\frac{3}{14} = 4\frac{3}{14}$

$\quad + 2\frac{5}{7} = + 2\frac{10}{14}$

$\quad\quad\quad\quad\quad\quad 6\frac{13}{14}$

14. $\quad 38\frac{3}{5} = 38\frac{6}{10}$

$\quad + 14\frac{7}{10} = + 14\frac{7}{10}$

$\quad\quad\quad\quad\quad\quad 52\frac{13}{10}$

$= 52 + 1\frac{3}{10}$

$= 53\frac{3}{10}$

15. $\quad 9\frac{7}{8} = 9\frac{7}{8}$

$\quad - 2\frac{1}{4} = - 2\frac{2}{8}$

$\quad\quad\quad\quad\quad\quad 7\frac{5}{8}$

16. $\quad 13\frac{1}{6} = 13\frac{1}{6}$

$\quad -12\frac{2}{3} = -12\frac{4}{6}$

$13\frac{1}{6} = 12 + 1\frac{1}{6} = 12 + \frac{7}{6}$

$\quad\quad\quad\quad 12\frac{7}{6}$

$\quad\quad\quad -12\frac{4}{6}$

$\quad\quad\quad\quad\quad \frac{3}{6} = \frac{1}{2}$

17. $\quad 19 = 18\frac{10}{10}$

$\quad - 18\frac{9}{10} = - 18\frac{9}{10}$

$\quad\quad\quad\quad\quad\quad\quad \frac{1}{10}$

18. $\quad 53\frac{1}{5} = 53\frac{18}{90}$

$\quad 16\frac{3}{10} = 16\frac{27}{90}$

$\quad + \frac{8}{9} = + \frac{80}{90}$

$\quad\quad\quad\quad\quad 69\frac{125}{90}$

$= 69 + 1\frac{35}{90}$

$= 70\frac{7}{18}$

19. 1st weight + 2nd weight = total weight

$\quad 2\frac{3}{5} \quad + \quad 3\frac{1}{4} \quad$ = total weight

$\quad 2\frac{3}{5} = 2\frac{12}{20}$

$\quad + 3\frac{1}{4} = 3\frac{5}{20}$

$\quad\quad\quad\quad\quad 5\frac{17}{20}$

The total weight is $5\frac{17}{20}$ pounds.

20. Let x = fraction owned by other partner

Carrie's part	+	John's part	+	partner's part	=	whole business
$\frac{3}{7}$	+	$\frac{1}{6}$	+	x	=	1

$$42\left(\frac{3}{7} + \frac{1}{6} + x\right) = 42(1)$$

$$42\left(\frac{3}{7}\right) + 42\left(\frac{1}{6}\right) + 42(x) = 42(1)$$

$$18 + 7 + 42x = 42$$
$$25 + 42x = 42$$
$$25 + 42x + (-25) = 42 + (-25)$$
$$42x = 17$$

$$\frac{42x}{42} = \frac{17}{42}$$

$$x = \frac{17}{42}$$

The other partner owns $\frac{17}{42}$ of the business.

CHAPTER 4 SOLUTIONS TO PRACTICE TEST

1. $\frac{5}{5} \div \frac{2}{5} = \frac{5}{5} \cdot \frac{5}{2}$

$$= \frac{5 \cdot 5}{5 \cdot 2}$$
$$= \frac{5}{2}$$

2. $-\frac{5}{7} \cdot \frac{9}{5} = -\frac{5 \cdot 9}{7 \cdot 5}$

$$= -\frac{9}{7}$$

3. $\frac{3x}{8} + \frac{x}{8} = \frac{3x + x}{8}$

$$= \frac{4x}{8}$$
$$= \frac{4x}{4 \cdot 2}$$
$$= \frac{x}{2}$$

4. $\frac{1}{10} - \frac{2}{x} = \frac{1 \cdot x}{10 \cdot x} - \frac{2 \cdot 10}{x \cdot 10}$

$$= \frac{x}{10x} - \frac{20}{10x}$$
$$= \frac{x - 20}{10x}$$

5. $\frac{x^2 y}{z^3} \cdot \frac{z}{xy^2} = \frac{x \cdot x \cdot y \cdot z}{z \cdot z \cdot z \cdot x \cdot y \cdot y}$

$$= \frac{x}{yz^2}$$

6. $-\frac{4}{9} \cdot -\frac{16}{30} = \frac{2 \cdot 2 \cdot 2 \cdot 2 \cdot 2 \cdot 2}{3 \cdot 3 \cdot 2 \cdot 3 \cdot 5}$

$$= \frac{32}{135}$$

7. $\frac{8b}{11} + \frac{3}{22} = \frac{8b \cdot 2}{11 \cdot 2} + \frac{3}{22}$

$$= \frac{16b}{22} + \frac{3}{22}$$
$$= \frac{16b + 3}{22}$$

8. $-\frac{4}{13m} - \frac{6}{13m} = \frac{-4 - 6}{13m}$

$$= -\frac{10}{13m}$$

9. $10x^2 \div \frac{x}{12} = \frac{10x^2}{1} \cdot \frac{12}{x}$

$$= \frac{10x \cdot x \cdot 12}{1 \cdot x}$$
$$= 120x$$

10. $3\frac{1}{8} \div \frac{5}{24} = \frac{25}{8} \div \frac{5}{24}$

$$= \frac{25}{8} \cdot \frac{24}{5}$$
$$= \frac{5 \cdot 5 \cdot 3 \cdot 8}{8 \cdot 5}$$
$$= 15$$

11.
$$5\frac{1}{8} = 5\frac{1}{8}$$
$$3\frac{1}{2} = 3\frac{4}{8}$$
$$+ 7\frac{3}{4} = +7\frac{6}{8}$$
$$15\frac{11}{8}$$
$$= 15 + 1\frac{3}{8}$$
$$= 16\frac{3}{8}$$

12.
$$\frac{4x}{7} \cdot \frac{14}{8x^4} = \frac{4 \cdot x \cdot 2 \cdot 7}{7 \cdot 4 \cdot 2 \cdot x \cdot x \cdot x \cdot x}$$
$$= \frac{1}{x^3}$$

13.
$$-\frac{15}{2} \div -\frac{45}{4} = -\frac{15}{2} \cdot -\frac{4}{45}$$
$$= \frac{15 \cdot 2 \cdot 2}{2 \cdot 3 \cdot 15}$$
$$= \frac{2}{3}$$

14.
$$9\frac{1}{5} \cdot 4\frac{3}{10} = \frac{46}{5} \cdot \frac{43}{10}$$
$$= \frac{2 \cdot 23 \cdot 43}{5 \cdot 2 \cdot 5}$$
$$= \frac{989}{25}$$
$$\text{or } 39\frac{14}{25}$$

15.
$$20 \div 5\frac{3}{4} = 20 \div \frac{23}{4}$$
$$= \frac{20}{1} \cdot \frac{4}{23}$$
$$= \frac{80}{23}$$
$$\text{or } 3\frac{11}{23}$$

16.
$$\left(\frac{12}{7} \cdot \frac{21}{6}\right) \div 8 = \left(\frac{2 \cdot 2 \cdot 3 \cdot 3 \cdot 7}{7 \cdot 2 \cdot 3}\right) \div 8$$
$$= 6 \div 8$$
$$= \frac{6}{8} = \frac{3}{4}$$

17.
$$\frac{5}{6} - \frac{2}{3} + \frac{11}{12} = \frac{5}{6} - \frac{4}{6} + \frac{11}{12}$$
$$= \frac{1}{6} + \frac{11}{12}$$
$$= \frac{2}{12} + \frac{11}{12}$$
$$= \frac{13}{12}$$

18.
$$\frac{\frac{4x}{15}}{\frac{16x^2}{60}} = \frac{4x}{15} \div \frac{16x^2}{60}$$
$$= \frac{4x}{15} \cdot \frac{60}{16x^2}$$
$$= \frac{4 \cdot x \cdot 4 \cdot 15}{15 \cdot 4 \cdot 4 \cdot x \cdot x}$$
$$= \frac{1}{x}$$

19.
$$\frac{4 + \frac{3}{8}}{5 - \frac{1}{4}} = \frac{8\left(4 + \frac{3}{8}\right)}{8\left(5 - \frac{1}{4}\right)}$$
$$= \frac{8(4) + 8\left(\frac{3}{8}\right)}{8(5) - 8\left(\frac{1}{4}\right)}$$
$$= \frac{32 + 3}{40 - 2}$$
$$= \frac{35}{38}$$

20.

$$-\frac{7}{8}x = \frac{5}{16}$$

$$-\frac{8}{7}\left(-\frac{7}{8}x\right) = -\frac{8}{7}\left(\frac{5}{16}\right)$$

$$x = -\frac{8 \cdot 5}{7 \cdot 2 \cdot 8}$$

$$x = -\frac{5}{14}$$

21.

$$\frac{x}{4} + x = -\frac{25}{16}$$

$$16\left(\frac{x}{4} + x\right) = 16\left(-\frac{25}{16}\right)$$

$$16\left(\frac{x}{4}\right) + 16(x) = 16\left(-\frac{25}{16}\right)$$

$$4x + 16x = -25$$
$$20x = -25$$

$$\frac{20x}{20} = \frac{-25}{20}$$

$$x = -\frac{5}{4}$$

22.

$$\frac{1}{4} + \frac{x}{3} = \frac{5}{6} + \frac{x}{2}$$

$$12\left(\frac{1}{4} + \frac{x}{3}\right) = 12\left(\frac{5}{6} + \frac{x}{2}\right)$$

$$12\left(\frac{1}{4}\right) + 12\left(\frac{x}{3}\right) = 12\left(\frac{5}{6}\right) + 12\left(\frac{x}{2}\right)$$

$$3 + 4x = 10 + 6x$$
$$3 + 4x - 4x = 10 + 6x - 4x$$
$$3 = 10 + 2x$$
$$3 - 10 = 10 + 2x - 10$$
$$-7 = 2x$$

$$\frac{-7}{2} = \frac{2x}{2}$$

$$-\frac{7}{2} = x$$

23. $-6x = -6\left(-\frac{7}{12}\right)$

$$= -\frac{6}{1} \cdot -\frac{7}{12}$$

$$= \frac{6 \cdot 7}{1 \cdot 2 \cdot 6}$$

$$= \frac{7}{2}$$

24. $xy = \frac{2}{3} \cdot 4\frac{5}{6}$

$$= \frac{2}{3} \cdot \frac{29}{6}$$

$$= \frac{2 \cdot 29}{3 \cdot 2 \cdot 3}$$

$$= \frac{29}{9}$$

or $3\frac{2}{9}$

25.

original length	−	length removed	=	length of remaining piece
$12\frac{2}{5}$	−	$5\frac{7}{10}$	=	remaining length

$$12\frac{2}{5} = 12\frac{4}{10}$$
$$-\ 5\frac{7}{10} = -5\frac{7}{10}$$

$$12\frac{4}{10} = 11 + 1\frac{4}{10}$$

$$= 11 + \frac{14}{10}$$

$$= 11\frac{14}{10}$$

$$11\frac{14}{10}$$
$$-\ 5\frac{7}{10}$$
$$\overline{\ \ 6\frac{7}{10}\ \ }$$

He has $6\frac{7}{10}$ feet of rope remaining.

165

5.1 SOLUTIONS TO EXERCISES

1. 8.07
   eight and seven hundredths

2. 142.5
   one hundred forty-two and five tenths

3. 16.341
   sixteen and three hundred forty-one thousandths

4. 12.4

5. 57.028
      ↑
   thousandths place

6. 189.14

7. $0.6 = \dfrac{6}{10} = \dfrac{3}{5}$

8. $3.72 = 3\dfrac{72}{100} = 3\dfrac{18}{25}$

9. $0.045 = \dfrac{45}{1000} = \dfrac{9}{200}$

10. $14.606 = 14\dfrac{606}{1000} = 14\dfrac{303}{500}$

11. $591.44 = 591\dfrac{44}{100} = 591\dfrac{11}{25}$

12. $0.3005 = \dfrac{3005}{10000} = \dfrac{601}{2000}$

13. $0.21 \;\; < \;\; 0.22$

14. $161.351 \;\; > \;\; 161.349$

15. $25000 \;\; > \;\; 0.000025$

16. $0.13000 = 0.130$

17. 0.49 rounded to the nearest tenth is 0.5.

   The digit to the right of the tenths place is greater than or equal to 5, so add 1 to the tenths place and drop all digits to the right of the tenths place.

18. 2.4631 rounded to the nearest hundredth is 2.46.

The digit to the right of the hundredths place is less than 5, so leave the hundredths place as is and drop all digits to the right of the hundredths place.

19. 0.5786 rounded to the nearest thousandth is 0.579

   The digit to the right of the thousandths place is greater than or equal to 5, so add 1 to the thousandths place and drop all digits to the right of the thousandths place.

20. 42,301.89 rounded to the nearest ten is 42,300.

   The digit to the right of the tens place is less than 5, so leave the tens place as is and change the digits to the right to 0s and drop the digits to the right of the decimal.

5.2 SOLUTIONS TO EXERCISES

1. $\begin{array}{r} 3.6 \\ +\ 9.2 \\ \hline 12.8 \end{array}$

2. $\begin{array}{r} {}^{1}\ \ \ {}^{1}\ \ \ \\ 14.810 \\ +\ 16.097 \\ \hline 30.907 \end{array}$

3. $\begin{array}{r} {}^{9} \\ {}^{6}\ \cancel{\rlap{\,}0}\ {}^{11} \\ 7.01 \\ -\ 6.94 \\ \hline 0.07 \end{array}$

   $-7.01 + 6.94 = -0.07$

4. $\begin{array}{r} {}^{1}\ \ \ \ \ \\ 39.113 \\ +\ 23.040 \\ \hline 62.153 \end{array}$

   $-39.113 + (-23.04) = -62.153$

5.

$$\begin{array}{r} \scriptstyle 9 \\ \scriptstyle 3\ \cancel{10}\ 10 \\ 14.\cancel{0}\cancel{0} \\ -\ 0.33 \\ \hline 13.67 \end{array}$$

6. $-9.246 - 10.0351 = -9.246 + (-10.0351)$

$$\begin{array}{r} \scriptstyle 1 \\ 9.2460 \\ +\ 10.0351 \\ \hline 19.2811 \end{array}$$

$-9.246 - 10.0351 = -19.2811$

7.

$$\begin{array}{r} \scriptstyle 12\ 10 \\ \scriptstyle 8\cancel{3}\ \cancel{3}\cancel{1}16 \\ 59\cancel{3}.1\cancel{6} \\ -\ 49.58 \\ \hline 543.58 \end{array}$$

8.

$$\begin{array}{r} \scriptstyle 11 \\ \scriptstyle 1\ \cancel{2}\cancel{1}10 \\ 22.\cancel{0} \\ -\ 4.9 \\ \hline 17.1 \end{array}$$

9.

$$\begin{array}{r} \scriptstyle 710 \\ 647.8\cancel{0} \\ -\ 37.77 \\ \hline 610.03 \end{array}$$

10.

$$\begin{array}{r} \scriptstyle 10 \\ \scriptstyle 1\ \cancel{1}\ 16 \\ 21.\cancel{6}47 \\ -19.840 \\ \hline 1.807 \end{array}$$

$21.647 + (-19.84) = 1.807$

11. $x - z = 25 - 0.106$
$= 24.894$

12. $x + y + z = 25 + 3.7 + 0.106$
$= 28.806$

13. $z - y = 0.106 - 3.7$
$= -3.594$

14. $x + 7.3 = 9.8$
$2.6 + 7.3 = 9.8$
$9.9 = 9.8$

False

$x = 2.6$ is not a solution.

15. $51.7 - y = 92$
$51.7 - (-40.3) = 92$
$92 = 92$
True

-40.3 is a solution.

16. $2.4 - m = m + 0.3$
$2.4 - 1.05 = 1.05 + 0.3$
$1.35 = 1.35$
True

1.05 is a solution.

17. $41.8x + 13.5 - 20.3x - 18.6$
$= 41.8x - 20.3x + 13.5 - 18.6$
$= 21.5x - 5.1$

18. $-7.81y - 5.68 - 10.06y + 7.14$
$= -7.81y - 10.06y - 5.68 + 7.14$
$= -17.87y + 1.46$

19. price   -   price   =   price change
2nd day   1st day

$\$1.359 \quad - \quad \$1.339 \quad = \quad \$0.02$

The price changed by $0.02.

20. $P = 2l + 2w$
$= 2(38.4 \text{ feet}) + 2(22.6 \text{ feet})$
$= 76.8 \text{ feet} + 45.2 \text{ feet}$
$= 122 \text{ feet}$

The perimeter is 122 feet.

5.3 SOLUTIONS TO EXERCISES

1.

$$\begin{array}{r} 0.5 \\ \times\ 0.3 \\ \hline .15 \end{array}$$

2.   1.68
    $\times$ 0.73
        504
    1176
    1.2264

3.   5.78
    $\times$ 4.1
       578
    2312
    23.698

4. 1.0006
    $\times$ 4.2
    20012
    40024
    4.20252

5. $(-1.268)(100) = -126.8$

    2 zeros in 100, so move decimal point
    2 places to the right.

6. $(-9.07)(3.5) = -31.745$

7. $(-10.206)(-0.01) = 0.10206$

    2 decimal places, so move decimal point 2
    places to the left.

8. $(561.0704)(1000) = 561{,}070.4$

    3 zeros in 1000, so move decimal point
    3 places to the right.

9.   2.0056
    $\times$ 7.9
    180504
    140392
    15.84424

10. 3.1154
    $\times$ 0.8
    2.49232

11. $xy = (35)(9.4)$
    $= 329$

12. $xy = (-4.7)(-6)$
    $= 28.2$

13. $xy = (-23.1)(5.02)$
    $= -115.962$

14. $0.9x = 5.13$
    $0.9(5.7) = 5.13$
       $5.13 = 5.13$
         True

    5.7 is a solution.

15. $-0.3x = -10.884$
    $-0.3(-36.28) = -10.884$
        $10.884 = -10.884$
          False

    -36.28 is not a solution.

16. $C = 2\pi r$
    $= 2\pi(3 \text{ feet})$
    $= 6\pi \text{ feet}$

    $6\pi \text{ feet} \approx 6(3.14) \text{ feet}$
    $= 18.84 \text{ feet}$

17. $C = 2\pi r$
    $= 2\pi(8.4 \text{ meters})$
    $= 16.8\pi \text{ meters}$

    $16.8\pi \text{ meters} \approx 16.8(3.14) \text{ meters}$
    $= 52.752 \text{ meters}$

18. $C = \pi d$
    $= \pi(12 \text{ yards})$
    $= 12\pi \text{ yards}$

    $12\pi \text{ yards} \approx 12(3.14) \text{ yards}$
    $= 37.68 \text{ yards}$

19. | number of gallons | $\cdot$ | price per gallon | = | total cost |
    |---|---|---|---|---|
    | 2 | $\cdot$ | $15.96 | = | $31.92 |

    The total cost is $31.92.

20. | number of yards | $\cdot$ | price per yard | = | total cost |
    |---|---|---|---|---|
    | 2.3 | $\cdot$ | $9.98 | = | $22.954 |

    The total cost is $22.95.

5.4 SOLUTIONS TO EXERCISES

1.
```
    0.205
4)0.820
    8
    2
    0
    20
    20
    0
```

2.
```
      600.
0.08.)48.00.
      48
      0
      0
      0
      0
      0
```
$-48 \div 0.08 = -600$

3.
```
      9.6
0.78.)7.48.8
      7 02
      468
      468
      0
```

4.
```
      8.4
0.33.)2.77.2
      2 64
      132
      132
      0
```
$2.772 \div (-0.33) = -8.4$

5.
```
    1.604
3)4.812
    3
    18
    18
    1
    0
    12
    12
    0
```

6.
```
      500.
0.16.)80.00.
      80
```

7.
```
       5000.
0.009.)45.000.
       45
```

8. $(-3.6) \div (-10,000) = -0.00036$
4 zeros, move decimal point 4 places to the left for division.

9.
```
       590.
0.055.)32.450.
       275
       495
       495
       0
       0
       0
```
$\dfrac{32.45}{-0.055} = -590$

10.
```
       72.
0.081.)5.832.
       567
       162
       162
       0
```

11.
```
       1220.428
0.07.)85.43.000
       7
       15
       14
       14
       14
       3
       0
       30
       28
       20
       14
       60
       56
       4
```
$\dfrac{85.43}{0.07} \approx 1220.43$

$$\begin{array}{r} 704.076 \\ \end{array}$$

12. $0.13.\overline{)91.53.000}$

$$\begin{array}{r} 91 \\ \hline 53 \\ 52 \\ \hline 10 \\ 0 \\ \hline 100 \\ 91 \\ \hline 90 \\ 78 \\ \hline 12 \end{array}$$

$$\frac{91.53}{0.13} \approx 704.08$$

13. $z \div y = 3.18 \div 0.5$
$$= 6.36$$

14. $x \div 0.002 = 4.62 \div 0.002$
$$= 2310$$

15. $\dfrac{x}{6} = 4.02$

$$\dfrac{24.12}{6} = 4.02$$

$$4.02 = 4.02$$
True
24.12 is a solution.

16. $\dfrac{y}{1000} = 0.31$

$$\dfrac{3100}{1000} = 0.31$$

$$3.1 = 0.31$$
False
3100 is not a solution.

17. Let x = the length

Area = $w \cdot l$
$$80.6 = 6.5 \cdot x$$

$$\dfrac{80.6}{6.5} = \dfrac{6.5 \cdot x}{6.5}$$

$$12.4 = x$$
The length is 12.4 meters.

18. Let x = number of meters

number of inches per meter	·	number of meters	=	total number of inches
39.37	·	x	=	300

$$39.37x = 300$$

$$\dfrac{39.37x}{39.37} = \dfrac{300}{39.37}$$

$$x = 7.6$$

There are 7.6 meters in 300 inches.

5.5 SOLUTIONS TO EXERCISES

	Actual	Estimate
1.	3.7	4
	9.2	9
	+ 4.5	+ 5
	17.4	18

	Actual	Estimate
2.	16.4	16
	− 8.9	− 9
	7.5	7

	Actual	Estimate
3.	512.3	500
	× 8	× 8
	4098.4	4000

Actual

$$\begin{array}{r} 27.558 \\ \end{array}$$

4. $8.5.\overline{)234.2.500}$

$$\begin{array}{r} 170 \\ \hline 642 \\ 595 \\ \hline 475 \\ 425 \\ \hline 500 \\ 425 \\ \hline 750 \\ 680 \\ \hline 70 \end{array}$$

$234.25 \div 8.5 = 27.56$

Estimate

$$9\overline{)230}$$
$$\begin{array}{r} 25 \\ 9\overline{)230} \\ \underline{18} \\ 50 \\ \underline{45} \\ 5 \end{array}$$

5. Actual

$(-6.3)^2 = (-6.3)(-6.3)$
$\qquad = 39.69$

Estimate

$(-6)(-6) = 36$

6.

Actual	Estimate
5.016	5
× 3.42	× 3
10032	15
20064	
15048	
17.15472	

7. $-2x + y = -2(3.4) + (-1.2)$
$\qquad = -6.8 + (-1.2)$
$\qquad = -8$

8. $z^2 = (-0.8)^2$
$\qquad = (-0.8)(-0.8)$
$\qquad = 0.64$

9. $\dfrac{xy}{z} = \dfrac{(3.4)(-1.2)}{-0.8}$

$\qquad = \dfrac{-4.08}{-0.8}$

$\qquad = 5.1$

10. $\quad 12x - 4.2 = 3$
$12(-0.6) - 4.2 = 3$
$\quad -7.2 - 4.2 = 3$
$\qquad\qquad -11.4 = 3$
$\qquad\qquad\quad$ False

-0.6 is not a solution.

11. $\quad 5x - 7.4 = 4x + 11.9$
$5(19.3) - 7.4 = 4(19.3) + 11.9$
$\quad 96.5 - 7.4 = 77.2 + 11.9$
$\qquad\qquad 89.1 = 89.1$
$\qquad\qquad\quad$ True

19.3 is a solution.

12. $\quad 10x + 8.2 = 11x - 9.1$
$10(0.9) + 8.2 = 11(0.9) - 9.1$
$\quad 9 + 8.2 = 9.9 - 9.1$
$\qquad 17.2 = 0.8$
$\qquad\quad$ False

0.9 is not a solution.

13. $(-4.9)^2 = (-4.9)(-4.9)$
$\qquad\quad = 24.01$

14. $(5.8 + 0.6)(9.7 - 1.2) = (6.4)(8.5)$
$\qquad\qquad\qquad\qquad = 54.4$

15. $\dfrac{0.813 - 4.62}{0.03} = \dfrac{-3.807}{0.03}$

$\qquad\qquad\qquad = -126.9$

16. $\dfrac{9 + 0.81}{18} = \dfrac{9.81}{18}$

$\qquad\qquad = 0.545$

17. $4.3(6.5 - 4.1) = 4.3(2.4)$
$\qquad\qquad\quad = 10.32$

18. $2.7 + (0.3)^2 = 2.7 + 0.09$
$\qquad\qquad\quad = 2.79$

19. $A = (\text{length}) \cdot (\text{width})$
$\quad \approx (21 \text{ feet}) \cdot (9 \text{ feet})$
$\quad = 189 \text{ sq. feet}$

The area is approximately 189 sq. feet.

20. $C = 2\pi r$
$\quad = 2(3.14)(5 \text{ inches})$
$\quad = 31.4 \text{ inches}$

The circumference is approximately 31.4 inches.

5.6 SOLUTIONS TO EXERCISES

1.
$$\begin{array}{r} 1.8 \\ 5\overline{)9.0} \\ \underline{5} \\ 40 \\ \underline{40} \\ 0 \end{array}$$

$$\frac{9}{5} = 1.8$$

2.
$$\begin{array}{r} 1.75 \\ 4\overline{)7.00} \\ \underline{4} \\ 30 \\ \underline{28} \\ 20 \\ \underline{20} \\ 0 \end{array}$$

$$\frac{7}{4} = 1.75$$

3.
$$\begin{array}{r} 0.45 \\ 20\overline{)9.00} \\ \underline{80} \\ 100 \\ \underline{100} \\ 0 \end{array}$$

$$\frac{9}{20} = 0.45$$

4.
$$\begin{array}{r} 0.416 \\ 12\overline{)5.000} \\ \underline{48} \\ 20 \\ \underline{12} \\ 80 \\ \underline{72} \\ 8 \end{array}$$

$$\frac{5}{12} \approx 0.42$$

5.
$$\begin{array}{r} 0.133 \\ 15\overline{)2.000} \\ \underline{15} \\ 50 \\ \underline{45} \\ 50 \\ \underline{45} \\ 5 \end{array}$$

$$\frac{2}{15} \approx 0.13$$

6.
$$\begin{array}{r} 0.692 \\ 13\overline{)9.000} \\ \underline{78} \\ 120 \\ \underline{117} \\ 30 \\ \underline{26} \\ 4 \end{array}$$

$$\frac{9}{13} \approx 0.69$$

7. $0.27 = \dfrac{27}{100}$

8. $0.425 = \dfrac{425}{1000}$

$= \dfrac{17}{40}$

9. $0.0008 = \dfrac{8}{10000}$

$= \dfrac{1}{1250}$

10. $19.25 = 19\dfrac{25}{100}$

$= 19\dfrac{1}{4}$

11. $5.74 = 5\dfrac{74}{100}$

$= 5\dfrac{37}{50}$

12. $3.444 = 3\dfrac{444}{1000}$

$= 3\dfrac{111}{250}$

13. $\dfrac{4}{9} = \dfrac{4 \cdot 8}{9 \cdot 8} = \dfrac{32}{72}$

$\dfrac{32}{72} > \dfrac{31}{72}$

$\dfrac{4}{9} > \dfrac{31}{72}$

14. $\dfrac{21}{23} = 0.913$

$\dfrac{41}{43} = 0.953$

$0.913 < 0.953$

$\dfrac{21}{23} < \dfrac{41}{43}$

15. $\dfrac{71}{12} = 5.917$

$5.917 < 5.92$

$\dfrac{71}{12} < 5.92$

16. $\dfrac{538}{19} = 28.316$

$28.316 < 29.476$

$\dfrac{538}{19} < 29.476$

17. $0.814, \ 0.830, \ 0.836$

18. Original numbers $\dfrac{11}{9}$ 1.22 $\dfrac{13}{8}$

     Decimals   1.222   1.220   1.625

     Compare in order 2nd 1st 3rd

     $1.22,\ \dfrac{11}{9},\ \dfrac{13}{8}$

19. $\dfrac{4}{3} - 5(2.6) = \dfrac{4}{3} - 13$

$$= \dfrac{4}{3} - \dfrac{39}{3}$$

$$= -\dfrac{35}{3}$$

20. $\dfrac{1}{8}(-10.2 - 21.8) = \dfrac{1}{8}(-32)$

$$= -4$$

5.7 SOLUTIONS TO EXERCISES

1. $x + 3.5 = 18.2$
   $x + 3.5 + (-3.5) = 18.2 + (-3.5)$
            $x = 14.7$

2. $y - 8.9 = 17.3$
   $y - 8.9 + 8.9 = 17.3 + 8.9$
          $y = 26.2$

3. $6y = 2.58$

   $\dfrac{6y}{6} = \dfrac{2.58}{6}$

     $y = 0.43$

4. $0.2m = 15.26$

   $\dfrac{0.2m}{0.2} = \dfrac{15.26}{0.2}$

     $m = 76.3$

5. $5x + 11.21 = 6x - 7$
  $5x + 11.21 + (-5x) = 6x - 7 + (-5x)$
          $11.21 = x - 7$
     $11.21 + 7 = x - 7 + 7$
          $18.21 = x$

6. $12y - 10.31 = 10y + 1.77$
  $12y - 10.31 + (-10y) = 10y + 1.77 + (-10y)$
       $2y - 10.31 = 1.77$
  $2y - 10.31 + 10.31 = 1.77 + 10.31$
          $2y = 12.08$

     $\dfrac{2y}{2} = \dfrac{12.08}{2}$

       $y = 6.04$

7. $4(x - 2.9) = 11.2$
     $4(x) - 4(2.9) = 11.2$
       $4x - 11.6 = 11.2$
  $4x - 11.6 + 11.6 = 11.2 + 11.6$
         $4x = 22.8$

     $\dfrac{4x}{4} = \dfrac{22.8}{4}$

       $x = 5.7$

8. $7(n + 3.3) = 87.5$
     $7(n) + 7(3.3) = 87.5$
       $7n + 23.1 = 87.5$
  $7n + 23.1 + (-23.1) = 87.5 + (-23.1)$
         $7n = 64.4$

     $\dfrac{7n}{7} = \dfrac{64.4}{7}$

       $n = 9.2$

9. $0.6x + 0.12 = -0.24$
    $100(0.6x + 0.12) = 100(-0.24)$
  $100(0.6x) + 100(0.12) = 100(-0.24)$
        $60x + 12 = -24$
   $60x + 12 + (-12) = -24 + (-12)$
          $60x = -36$

     $\dfrac{60x}{60} = \dfrac{-36}{60}$

       $x = -0.6$

10.
$$6x - 12.5 = x$$
$$10(6x - 12.5) = 10(x)$$
$$10(6x) - 10(12.5) = 10(x)$$
$$60x - 125 = 10x$$
$$60x - 125 + (-60x) = 10x + (-60x)$$
$$-125 = -50x$$

$$\frac{-125}{-50} = \frac{-50x}{-50}$$

$$2.5 = x$$

11.
$$3.8a + 7 - 1.2a = 22.6$$
$$10(3.8a + 7 - 1.2a) = 10(22.6)$$
$$10(3.8a) + 10(7) - 10(1.2a) = 10(22.6)$$
$$38a + 70 - 12a = 226$$
$$26a + 70 = 226$$
$$26a + 70 + (-70) = 226 + (-70)$$
$$26a = 156$$

$$\frac{26a}{26} = \frac{156}{26}$$

$$a = 6$$

12.
$$-0.005x = 29.65$$
$$1000(-0.005x) = 1000(29.65)$$
$$-5x = 29650$$

$$\frac{-5x}{-5} = \frac{29650}{-5}$$

$$x = -5930$$

13.
$$y + 15.04 = 11.2$$
$$100(y + 15.04) = 100(11.2)$$
$$100(y) + 100(15.04) = 100(11.2)$$
$$100y + 1504 = 1120$$
$$100y + 1504 + (-1504) = 1120 + (-1504)$$
$$100y = -384$$

$$\frac{100y}{100} = \frac{-384}{100}$$

$$y = -3.84$$

14.
$$300x - 0.74 = 200x + 0.9$$
$$100(300x - 0.74) = 100(200x + 0.9)$$
$$100(300x) - 100(0.74) = 100(200x) + 100(0.9)$$
$$30000x - 74 = 20000x + 90$$
$$30000x - 74 + 74 = 20000x + 90 + 74$$
$$30000x = 20000x + 164$$
$$30000x + (-20000x) = 20000x + 164 + (-20000x)$$
$$10000x = 164$$

$$\frac{10000x}{10000} = \frac{164}{10000}$$

$$x = 0.0164$$

15.
$$10(x - 9.4) = 25$$
$$10(x) - 10(9.4) = 10(25)$$
$$10x - 94 = 250$$
$$10x - 94 + 94 = 250 + 94$$
$$10x = 344$$

$$\frac{10x}{10} = \frac{344}{10}$$

$$x = 34.4$$

16.
$$12x + 8.6 = 4(2x - 6.1)$$
$$12x + 8.6 = 4(2x) - 4(6.1)$$
$$12x + 8.6 = 8x - 24.4$$
$$12x + 8.6 + (-8x) = 8x - 24.4 + (-8x)$$
$$4x + 8.6 = -24.4$$
$$4x + 8.6 + (-8.6) = -24.4 + (-8.6)$$
$$4x = -33$$

$$\frac{4x}{4} = \frac{-33}{4}$$

$$x = -8.25$$

17.
$$0.9x + 42.1 = x - 57.09$$
$$100(0.9x + 42.1) = 100(x - 57.09)$$
$$100(0.9x) + 100(42.1) = 100(x) - 100(57.09)$$
$$90x + 4210 = 100x - 5709$$
$$90x + 4210 + (-90x) = 100x - 5709 + (-90x)$$
$$4210 = 10x - 5709$$

$$4210 + 5709 = 10x - 5709 + 5709$$
$$9919 = 10x$$

$$\frac{9919}{10} = \frac{10x}{10}$$

$$991.9 = x$$

18. $$0.003x - 15 = 0.009$$
$$0.003x - 15 + 15 = 0.009 + 15$$
$$0.003x = 15.009$$

$$\frac{0.003x}{0.003} = \frac{15.009}{0.003}$$

$$x = 5003$$

5.8 SOLUTIONS TO EXERCISES

1. $\sqrt{196} = 14$ since $14 \cdot 14 = 196$

2. $\sqrt{0} = 0$ since $0 \cdot 0 = 0$

3. $\sqrt{225} = 15$ since $15 \cdot 15 = 225$

4. $\sqrt{\dfrac{16}{81}} = \dfrac{4}{9}$ since $\dfrac{4}{9} \cdot \dfrac{4}{9} = \dfrac{16}{81}$

5. $\sqrt{\dfrac{1}{121}} = \dfrac{1}{11}$ since $\dfrac{1}{11} \cdot \dfrac{1}{11} = \dfrac{1}{121}$

6. $\sqrt{\dfrac{144}{49}} = \dfrac{12}{7}$ since $\dfrac{12}{7} \cdot \dfrac{12}{7} = \dfrac{144}{49}$

7. $\sqrt{11} \approx 3.317$

8. $\sqrt{21} \approx 4.583$

9. $\sqrt{37} \approx 6.083$

10. $\sqrt{84} \approx 9.165$

11. $\sqrt{92} \approx 9.592$

12. $\sqrt{143} = 11.958$

13. $$a^2 + b^2 = c^2$$
$$18^2 + 24^2 = c^2$$
$$324 + 576 = c^2$$
$$900 = c^2$$
$$30 = c$$

14. $$a^2 + b^2 = c^2$$
$$7^2 + 9^2 = c^2$$
$$49 + 81 = c^2$$
$$130 = c^2$$
$$\sqrt{130} = c \quad \text{or} \quad c \approx 11.402$$

15. $$a^2 + b^2 = c^2$$
$$21^2 + 30^2 = c^2$$
$$441 + 900 = c^2$$
$$1341 = c^2$$
$$\sqrt{1341} = c \quad \text{or} \quad c \approx 36.620$$

16. $$a^2 + b^2 = c^2$$
$$15^2 + 36^2 = c^2$$
$$225 + 1296 = c^2$$
$$1521 = c^2$$
$$39 = c$$

17.

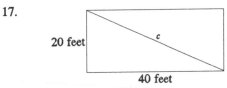

$$a^2 + b^2 = c^2$$
$$20^2 + 40^2 = c^2$$
$$400 + 1600 = c^2$$
$$2000 = c^2$$
$$\sqrt{2000} = c \quad \text{or} \quad c \approx 44.72$$

The diagonal is 44.72 feet.

18.

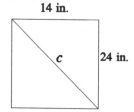

14 in.

c 24 in.

$$a^2 + b^2 = c^2$$
$$14^2 + 24^2 = c^2$$
$$196 + 576 = c^2$$
$$772 = c^2$$
$$\sqrt{772} = c \quad \text{or} \quad c \approx 27.78$$

The diagonal is 27.78 inches long.

CHAPTER 5 SOLUTIONS TO PRACTICE TEST

1. Eighty-four and seventeen hundredths

2. 268.7

3.
$$
\begin{array}{r}
{}^{2\ \ \,21} \\
5.716 \\
{}_{1}2.890 \\
+\ 12.398 \\
\hline
21.004
\end{array}
$$

4.
$$
\begin{array}{r}
{}^{1\ \ 1} \\
61.87 \\
+\ 4.76 \\
\hline
66.63
\end{array}
$$

$$-61.87 - 4.76 = -66.63$$

5.
$$
\begin{array}{r}
{}^{1\ 17\ 514} \\
27.64 \\
-\ 8.39 \\
\hline
19.25
\end{array}
$$

$$8.39 - 27.64 = -19.25$$

6.
$$
\begin{array}{r}
13.5 \\
5.26 \\
\hline
810 \\
270\ \ \\
675\ \ \ \ \\
\hline
71.010
\end{array}
$$

7.
$$
\begin{array}{r}
0.054 \\
0.71.\overline{)0.03.834} \\
3\,55 \\
\hline
284 \\
284 \\
\hline
0
\end{array}
$$

$$(-0.03834) \div (-0.71) = 0.054$$

8. 47.97<u>4</u>6 rounded to nearest hundredth is 47.97.

9. 0.147<u>8</u> rounded to nearest thousandth is 0.148.

10. 51.0<u>8</u>07 > 51.0<u>7</u>8

11. $\dfrac{3}{7} = 0.4286$

$$0.428\underline{6} < 0.429$$
$$\frac{3}{7} < 0.429$$

12. $0.625 = \dfrac{625}{1000}$

$$= \frac{5}{8}$$

13. $18.97 = 18\dfrac{97}{100}$

14.
$$
\begin{array}{r}
0.2 \\
85\overline{)17.0} \\
17\,0 \\
\hline
0
\end{array}
$$

15.
$$\begin{array}{r} 0.475 \\ 40\overline{)19.000} \\ \underline{160} \\ 300 \\ \underline{280} \\ 200 \\ \underline{200} \\ 0 \end{array}$$

$$\frac{19}{40} = 0.475$$

16. $(-0.4)^2 + 2.68 = 0.16 + 2.68$
$$= 2.84$$

17. $\dfrac{0.56 + 2.34}{-0.2} = \dfrac{2.9}{-0.2}$
$$= -14.5$$

18. $12.1x - 6.4 - 9.8x - 7.6$
$= 12.1x - 9.8x - 6.4 - 7.6$
$= 2.3x - 14$

19. $\sqrt{25} = 5$ since $5 \cdot 5 = 25$

20. $\sqrt{\dfrac{121}{36}} = \dfrac{11}{6}$ since $\dfrac{11}{6} \cdot \dfrac{11}{6} = \dfrac{121}{36}$

21. $\sqrt{175} \approx 13.229$ (calculator)

22. $a^2 + b^2 = c^2$
$7^2 + 7^2 = c^2$
$49 + 49 = c^2$
$\qquad 98 = c^2$
$\qquad \sqrt{98} = c$ or $c \approx 9.90$

The missing side is 9.90 inches.

23. $a^2 + b^2 = c^2$
$9^2 + 11^2 = c^2$
$81 + 121 = c^2$
$\qquad 202 = c^2$
$\qquad \sqrt{202} = c$ or $c \approx 14.21$

The missing side is 14.21 meters.

24.
$$0.6x + 5.7 = 0.3$$
$$0.6x + 5.7 + (-5.7) = 0.3 + (-5.7)$$
$$0.6x = -5.4$$
$$\frac{0.6x}{0.6} = \frac{-5.4}{0.6}$$
$$x = -9$$

25.
$$12(x + 1.6) = 10x - 8.4$$
$$12(x) + 12(1.6) = 10x - 8.4$$
$$12x + 19.2 = 10x - 8.4$$
$$12x + 19.2 + (-10x) = 10x - 8.4 + (-10x)$$
$$2x + 19.2 = -8.4$$
$$2x + 19.2 + (-19.2) = -8.4 + (-19.2)$$
$$2x = -27.6$$
$$\frac{2x}{2} = \frac{-27.6}{2}$$
$$x = -13.8$$

6.1 SOLUTIONS TO EXERCISES

1. $\dfrac{23}{31}$

2. $\dfrac{100}{53}$

3. $\dfrac{\frac{1}{3}}{8}$

4. $\dfrac{5\frac{4}{5}}{2\frac{1}{10}}$

5. $\dfrac{18}{57} = \dfrac{6}{19}$

6. $\dfrac{75}{375} = \dfrac{1}{5}$

7. $\dfrac{22 \text{ meters}}{162 \text{ meters}} = \dfrac{22}{162} = \dfrac{11 \cdot 2}{81 \cdot 2}$

   $\phantom{\dfrac{22 \text{ meters}}{162 \text{ meters}}} = \dfrac{11}{81}$

8. $\dfrac{80 \text{ feet}}{324 \text{ feet}} = \dfrac{80}{324} = \dfrac{20 \cdot 4}{81 \cdot 4}$

   $\phantom{\dfrac{80 \text{ feet}}{324 \text{ feet}}} = \dfrac{20}{81}$

9. $\dfrac{\$116}{\$20} = \dfrac{116}{20}$

   $\phantom{\dfrac{\$116}{\$20}} = \dfrac{4 \cdot 29}{4 \cdot 5}$

   $\phantom{\dfrac{\$116}{\$20}} = \dfrac{29}{5}$

10. $\dfrac{84 \text{ miles}}{105 \text{ miles}}$

    $= \dfrac{84}{105}$

    $= \dfrac{4 \cdot 21}{5 \cdot 21}$

    $= \dfrac{4}{5}$

11. $\dfrac{7.2}{103} = \dfrac{7.2 \cdot 10}{103 \cdot 10} = \dfrac{72}{1030}$

    $\phantom{\dfrac{7.2}{103}} = \dfrac{2 \cdot 36}{2 \cdot 515} = \dfrac{36}{515}$

12. $\dfrac{3.24}{5.28} = \dfrac{3.24 \cdot 100}{5.28 \cdot 100} = \dfrac{324}{528}$

    $\phantom{\dfrac{3.24}{5.28}} = \dfrac{12 \cdot 27}{12 \cdot 44} = \dfrac{27}{44}$

13. $\dfrac{14 \text{ women}}{18 \text{ men}} = \dfrac{7 \text{ women}}{9 \text{ men}}$

14. total people $= 18 + 14 = 32$

    $\dfrac{18 \text{ men}}{32 \text{ people}} = \dfrac{9 \text{ men}}{16 \text{ people}}$

15. $\dfrac{6 \text{ inches}}{20 \text{ inches}} = \dfrac{6}{20}$

    $= \dfrac{2 \cdot 3}{2 \cdot 10}$

    $= \dfrac{3}{10}$

16. Perimeter = distance around the rectangle
    $= 2(6 \text{ inches}) + 2(20 \text{ inches})$
    $= 52 \text{ inches}$

    $\dfrac{20 \text{ inches}}{52 \text{ inches}} = \dfrac{20}{52}$

    $= \dfrac{4 \cdot 5}{4 \cdot 13}$

    $= \dfrac{5}{13}$

6.2 SOLUTIONS TO EXERCISES

1. $\dfrac{10 \text{ lightpoles}}{2 \text{ miles}} = \dfrac{5 \text{ lightpoles}}{1 \text{ mile}}$

   or 5 lightpoles/mile

2. $\dfrac{36 \text{ computers}}{78 \text{ faculty members}} = \dfrac{6 \text{ computers}}{13 \text{ faculty members}}$

3. $\dfrac{9 \text{ defective lightbulbs}}{990 \text{ lightbulbs}} = \dfrac{1 \text{ defective lightbulb}}{110 \text{ lightbulbs}}$

4. $\dfrac{20 \text{ pear trees}}{220 \text{ feet}} = \dfrac{1 \text{ pear tree}}{11 \text{ feet}}$

5. $\dfrac{12 \text{ rooms}}{116 \text{ people}} = \dfrac{3 \text{ rooms}}{29 \text{ people}}$

6. $\dfrac{40 \text{ boxes}}{192 \text{ cookies}} = \dfrac{5 \text{ boxes}}{24 \text{ cookies}}$

7. $\dfrac{540 \text{ calories}}{5 \text{ ounces}} = \dfrac{108 \text{ calories}}{1 \text{ ounce}}$

   or 108 calories/ounce

8. $\dfrac{486 \text{ miles}}{9 \text{ hours}} = \dfrac{54 \text{ miles}}{1 \text{ hour}}$

   or 54 miles/hour

9. $\dfrac{\$2.24}{7 \text{ pears}} = \dfrac{\$0.32}{1 \text{ pear}}$

   or $0.32/pear

10. $\dfrac{432 \text{ miles}}{18 \text{ gallons}} = \dfrac{24 \text{ miles}}{1 \text{ gallon}}$

    or 24 miles/gallon

11. $\dfrac{\$7,000,000}{14 \text{ winners}} = \dfrac{\$500,000}{1 \text{ winner}}$

    or $500,000/winner

12. $\dfrac{1050 \text{ students}}{35 \text{ teachers}}$

    $= \dfrac{30 \text{ students}}{1 \text{ teacher}}$ or 30 students/teacher

13. $\dfrac{\$1.82}{16 \text{ ounces}} \approx \$0.11/\text{ounce}$

    $\dfrac{\$1.12}{9 \text{ ounces}} \approx \$0.12/\text{ounce}$

    $1.82 for 16 ounces is the better deal.

14. $\dfrac{\$1.00}{3 \text{ donuts}} \approx \$0.33/\text{donut}$

    $\dfrac{\$3.60}{12 \text{ donuts}} \approx \$0.30/\text{donut}$

    $3.60 for 12 donuts is the better deal.

15. $\dfrac{\$7.99}{20 \text{ ounces}} \approx \$0.40/\text{ounce}$

    $\dfrac{\$4.99}{12 \text{ ounces}} \approx \$0.42/\text{ounce}$

    $7.99 for 20 ounces is the better deal.

6.3 SOLUTIONS TO EXERCISES

1. $\dfrac{20 \text{ books}}{4 \text{ students}} = \dfrac{30 \text{ books}}{6 \text{ students}}$

2. $\dfrac{1\frac{1}{2} \text{ cups of flour}}{10 \text{ crepes}} = \dfrac{3\frac{3}{4} \text{ cups of flour}}{25 \text{ crepes}}$

3. $\dfrac{18 \text{ errors}}{12 \text{ pages}} = \dfrac{3 \text{ errors}}{2 \text{ pages}}$

4. $\dfrac{36 \text{ inches}}{3 \text{ feet}} = \dfrac{144 \text{ inches}}{12 \text{ feet}}$

5. $\dfrac{16}{12} = \dfrac{40}{32}$

$16 \cdot 32 = 12 \cdot 40$
$512 = 480$
False

The proportion is not true.

6. $\dfrac{40}{60} = \dfrac{1200}{1800}$

$40 \cdot 1800 = 60 \cdot 1200$
$72000 = 72000$
True

The proportion is true.

7. $\dfrac{0.9}{0.2} = \dfrac{4.5}{1.0}$

$0.9 \cdot 1.0 = 0.2 \cdot 4.5$
$0.9 = 0.9$
True

The proportion is true.

8. $\dfrac{2.8}{1.7} = \dfrac{14}{9}$

$2.8 \cdot 9 = 1.7 \cdot 14$
$25.2 = 23.8$
False

The proportion is not true.

9. $\dfrac{\frac{3}{5}}{\frac{7}{10}} = \dfrac{\frac{2}{5}}{\frac{7}{15}}$

$\dfrac{3}{5} \cdot \dfrac{7}{15} = \dfrac{7}{10} \cdot \dfrac{2}{5}$

$\dfrac{7}{25} = \dfrac{7}{25}$
True

The proportion is true.

10. $\dfrac{4\frac{1}{8}}{\frac{7}{3}} = \dfrac{24\frac{3}{4}}{14}$

$4\dfrac{1}{8} \cdot 14 = \dfrac{7}{3} \cdot 24\dfrac{3}{4}$

$\dfrac{33}{8} \cdot \dfrac{14}{1} = \dfrac{7}{3} \cdot \dfrac{99}{4}$

$\dfrac{231}{4} = \dfrac{231}{4}$
True

The proportion is true.

11. $\dfrac{x}{4} = \dfrac{110}{44}$

$x \cdot 44 = 4 \cdot 110$
$44x = 440$

$\dfrac{44x}{44} = \dfrac{440}{44}$

$x = 10$

12. $\dfrac{7}{y} = \dfrac{84}{132}$

$7 \cdot 132 = y \cdot 84$
$924 = 84y$

$\dfrac{924}{84} = \dfrac{84y}{84}$

$11 = y$

13. $\dfrac{25}{80} = \dfrac{z}{16}$

$25 \cdot 16 = 80 \cdot z$
$400 = 80z$

$\dfrac{400}{80} = \dfrac{80z}{80}$

$5 = z$

14. $\dfrac{7}{8} = \dfrac{98}{n}$

$7 \cdot n = 8 \cdot 98$

$7n = 784$

$\dfrac{7n}{7} = \dfrac{784}{7}$

$n = 112$

15. $\dfrac{\frac{3}{7}}{21} = \dfrac{x}{7}$

$\dfrac{3}{7} \cdot 7 = 21 \cdot x$

$3 = 21x$

$\dfrac{3}{21} = \dfrac{21x}{21}$

$\dfrac{1}{7} = x$

16. $\dfrac{9.4}{3.2} = \dfrac{4.7}{y}$

$9.4 \cdot y = 3.2 \cdot 4.7$

$9.4y = 15.04$

$\dfrac{9.4y}{9.4} = \dfrac{15.04}{9.4}$

$y = 1.6$

17. $\dfrac{\frac{8}{9}}{\frac{26}{27}} = \dfrac{2\frac{2}{3}}{z}$

$\dfrac{8}{9} \cdot z = \dfrac{26}{27} \cdot 2\dfrac{2}{3}$

$\dfrac{8}{9}z = \dfrac{26}{27} \cdot \dfrac{8}{3}$

$\dfrac{8}{9}z = \dfrac{208}{81}$

$\dfrac{9}{8} \cdot \dfrac{8}{9}z = \dfrac{9}{8} \cdot \dfrac{208}{81}$

$z = \dfrac{26}{9}$

18. $\dfrac{0.6}{y} = \dfrac{12}{400}$

$0.6 \cdot 400 = y \cdot 12$

$240 = 12y$

$\dfrac{240}{12} = \dfrac{12y}{12}$

$20 = y$

19. $\dfrac{x}{6.12} = \dfrac{0.91}{0.07}$

$x \cdot 0.07 = 6.12 \cdot 0.91$

$0.07x = 5.5692$

$\dfrac{0.07x}{0.07} = \dfrac{5.5692}{0.07}$

$x = 79.6$

20. $\dfrac{2036}{5694} = \dfrac{3122}{y}$

$2036 \cdot y = 5694 \cdot 3122$

$2036y = 17776668$

$\dfrac{2036y}{2036} = \dfrac{17776668}{2036}$

$y = 8731.17$

6.4 SOLUTIONS TO EXERCISES

1. Let x = number of completions

$\dfrac{5 \text{ completions}}{12 \text{ attempts}} = \dfrac{x \text{ completions}}{36 \text{ attempts}}$

$5 \cdot 36 = 12 \cdot x$

$180 = 12x$

$\dfrac{180}{12} = \dfrac{12x}{12}$

$15 = x$

He would have completed 15 passes.

2. Let x = number of attempts

$\dfrac{5 \text{ completions}}{12 \text{ attempts}} = \dfrac{25 \text{ completions}}{x \text{ attempts}}$

$5 \cdot x = 12 \cdot 25$

$5x = 300$

$\dfrac{5x}{5} = \dfrac{300}{5}$

$x = 60$

He would have attempted 60 passes.

3. Let x = wall's length

$\dfrac{1 \text{ inch}}{6 \text{ feet}} = \dfrac{3\frac{1}{3} \text{ inches}}{x \text{ feet}}$

$1 \cdot x = 6 \cdot 3\dfrac{1}{3}$

$x = 6 \cdot \dfrac{10}{3}$

$x = 20$

The wall's length is 20 feet.

4. Let x = wall's blueprint measurement

$\dfrac{1 \text{ inch}}{6 \text{ feet}} = \dfrac{x \text{ inches}}{48 \text{ feet}}$

$1 \cdot 48 = 6 \cdot x$

$48 = 6x$

$\dfrac{48}{6} = \dfrac{6x}{6}$

$8 = x$

The blueprint measurement is 8 inches.

5. Let x = number of bags

Area = (length) · (width)
     = (320 ft)(210 ft)
     = 67200 sq. ft

$$\frac{1 \text{ bag}}{5000 \text{ sq. ft}} = \frac{x \text{ bags}}{67200 \text{ sq. ft}}$$

$1 \cdot 67200 = 5000 \cdot x$
$67200 = 5000x$

$$\frac{67200}{5000} = \frac{5000x}{5000}$$

$13.44 = x$

14 bags of fertilizer should be purchased.

6. Let x = number of bags

Area = (side)^2
     = (220 ft)^2
     = 48400 sq. ft

$$\frac{1 \text{ bag}}{5000 \text{ sq. ft}} = \frac{x \text{ bags}}{48400 \text{ sq. ft}}$$

$1 \cdot 48400 = 5000 \cdot x$
$48400 = 5000x$

$$\frac{48400}{5000} = \frac{5000x}{5000}$$

$9.68 = x$

10 bags of fertilizer should be purchased.

7. Let x = value of home

$$\frac{\$1.65 \text{ tax}}{\$100 \text{ house value}} = \frac{\$3465 \text{ tax}}{x \text{ house value}}$$

$1.65 \cdot x = 100 \cdot 3465$
$1.65x = 346500$

$$\frac{1.65x}{1.65} = \frac{346500}{1.65}$$

$x = 210000$

Her house value is $210,000.

8. Let x = amount of tax

$$\frac{\$1.65 \text{ tax}}{\$100 \text{ house value}} = \frac{x \text{ tax}}{\$158,000 \text{ house value}}$$

$1.65 \cdot 158,000 = 100 \cdot x$
$260700 = 100x$

$$\frac{260700}{100} = \frac{100x}{100}$$

$2607 = x$

The property taxes are $2607.

9. Let x = number of hits

$$\frac{4 \text{ hits}}{9 \text{ times at bat}} = \frac{x \text{ hits}}{36 \text{ times at bat}}$$

$4 \cdot 36 = 9 \cdot x$
$144 = 9x$

$$\frac{144}{9} = \frac{9x}{9}$$

$16 = x$

He would be expected to make 16 hits.

10. Let x = number of times at bat

$$\frac{4 \text{ hits}}{9 \text{ times at bat}} = \frac{12 \text{ hits}}{x \text{ times at bat}}$$

$4 \cdot x = 9 \cdot 12$
$4x = 108$

$$\frac{4x}{4} = \frac{108}{4}$$

$x = 27$

He would have been at bat 27 times.

11. Let x = number of people preferring cake

$$\frac{5 \text{ prefer cake}}{7 \text{ people}} = \frac{x \text{ prefer cake}}{84 \text{ people}}$$

$$5 \cdot 84 = 7 \cdot x$$
$$420 = 7x$$

$$\frac{420}{7} = \frac{7x}{7}$$

$$60 = x$$

60 people prefer cake.

12. $7 - 5 = 2$ people out of 7 prefer pie

Let x = number of people preferring pie

$$\frac{2 \text{ prefer pie}}{7 \text{ people}} = \frac{x \text{ prefer pie}}{35 \text{ people}}$$

$$2 \cdot 35 = 7 \cdot x$$
$$70 = 7x$$

$$\frac{70}{7} = \frac{7x}{7}$$

$$10 = x$$

10 of the students are likely to prefer pie.

13. Let x = number of miles

$$\frac{16 \text{ gallons}}{400 \text{ miles}} = \frac{6 \text{ gallons}}{x \text{ miles}}$$

$$16 \cdot x = 400 \cdot 6$$
$$16x = 2400$$

$$\frac{16x}{16} = \frac{2400}{16}$$

$$x = 150$$

She can go 150 miles.

14. Let x = number of gallons

$$\frac{16 \text{ gallons}}{400 \text{ miles}} = \frac{x \text{ gallons}}{1560 \text{ miles}}$$

$$16 \cdot 1560 = 400 \cdot x$$
$$24960 = 400x$$

$$\frac{24960}{400} = \frac{400x}{400}$$

$$62.4 = x$$

He can expect to burn approximately 62 gallons of gas.

6.5 SOLUTIONS TO EXERCISES

1. ratio of pair of corresponding sides = $\frac{14}{7} = \frac{2}{1}$

2. ratio of pair of corresponding sides = $\frac{2}{6} = \frac{1}{3}$

3. $$\frac{14}{7} = \frac{26}{x}$$

$$14 \cdot x = 7 \cdot 26$$
$$14x = 182$$

$$\frac{14x}{14} = \frac{182}{14}$$

$$x = 13$$

The unknown side is of length 13.

4. $$\frac{36}{9} = \frac{42}{x}$$

$$36 \cdot x = 9 \cdot 42$$
$$36x = 378$$

$$\frac{36x}{36} = \frac{378}{36}$$

$$x = 10.5$$

The unknown side is of length 10.5.

5. $\dfrac{16}{56} = \dfrac{6}{x}$

$16 \cdot x = 56 \cdot 6$
$16x = 336$

$\dfrac{16x}{16} = \dfrac{336}{16}$

$x = 21$

The unknown side is of length 21.

6. $\dfrac{12.5}{2.5} = \dfrac{30.5}{x}$

$12.5 \cdot x = 2.5 \cdot 30.5$
$12.5x = 76.25$

$\dfrac{12.5x}{12.5} = \dfrac{76.25}{12.5}$

$x = 6.1$

The unknown side is of length 6.1

7. $\dfrac{8.1}{3.2} = \dfrac{x}{3.2}$

$8.1 \cdot 3.2 = 3.2 \cdot x$

$\dfrac{8.1 \cdot 3.2}{3.2} = \dfrac{3.2 \cdot x}{3.2}$

$8.1 = x$

The unknown side is 8.1.

8. $\dfrac{5}{3} = \dfrac{20}{x}$

$5 \cdot x = 3 \cdot 20$
$5x = 60$

$\dfrac{5x}{5} = \dfrac{60}{5}$

$x = 12$

The unknown side is 12.

9.   Let x = shadow's length

$\dfrac{40\text{-foot tree}}{22\text{-foot shadow}} = \dfrac{52\text{-foot tree}}{x\text{-foot shadow}}$

$40 \cdot x = 22 \cdot 52$
$40x = 1144$

$\dfrac{40x}{40} = \dfrac{1144}{40}$

$x = 28.6$

The shadow is 28.6 feet.

10.   Let x = height of tree

$\dfrac{6\text{-foot height}}{5\text{-foot shadow}} = \dfrac{x\text{-foot height}}{54\text{-foot shadow}}$

$6 \cdot 54 = 5 \cdot x$
$324 = 5x$

$\dfrac{324}{5} = \dfrac{5x}{5}$

$64.8 = x$

The tree is 64.8 feet tall.

CHAPTER 6 PRACTICE TEST SOLUTIONS

1. $\dfrac{381 \text{ roses}}{574 \text{ roses}} = \dfrac{381}{574}$

2. $\dfrac{9\frac{1}{4} \text{ dollars}}{12\frac{3}{8} \text{ dollars}} = \dfrac{9\frac{1}{4}}{12\frac{3}{8}}$

3. $\dfrac{90}{27} = \dfrac{9 \cdot 10}{9 \cdot 3}$

      $= \dfrac{10}{3}$

4. $\dfrac{75}{125} = \dfrac{25 \cdot 3}{25 \cdot 5}$

      $= \dfrac{3}{5}$

5. $\dfrac{\$130}{\$80} = \dfrac{130}{80}$

$= \dfrac{13 \cdot 10}{8 \cdot 10}$

$= \dfrac{13}{8}$

6. $\dfrac{36 \text{ feet}}{150 \text{ feet}} = \dfrac{36}{150}$

$= \dfrac{6 \cdot 6}{6 \cdot 25}$

$= \dfrac{6}{25}$

7. $\dfrac{5 \text{ houses}}{3 \text{ miles}}$

8. $\dfrac{12 \text{ fax machines}}{92 \text{ faculty members}} = \dfrac{3 \text{ fax machines}}{23 \text{ faculty members}}$

9. $\dfrac{570 \text{ km}}{6 \text{ hr}} = 95 \text{ km/hr}$

10. $\dfrac{18 \text{ in.}}{30 \text{ days}} = 0.6 \text{ in./day}$

11. $\dfrac{\$1.49}{12 \text{ shells}} \approx \$0.124/\text{shell}$

$\dfrac{\$1.79}{18 \text{ shells}} \approx \$0.099/\text{shell}$

18 for $1.79 is the better buy.

12. $\dfrac{\$0.89}{3 \text{ lb}} \approx \$0.297/\text{lb}$

$\dfrac{\$1.99}{10 \text{ lb}} \approx \$0.199/\text{lb}$

10 pounds for $1.99 is the better buy.

13. $\dfrac{6 \text{ cups}}{2 \text{ bottles}} = \dfrac{24 \text{ cups}}{8 \text{ bottles}}$

14. $\dfrac{81 \text{ feet}}{27 \text{ yards}} = \dfrac{36 \text{ feet}}{12 \text{ yards}}$

15. $\dfrac{15}{75} = \dfrac{90}{450}$

$15 \cdot 450 = 75 \cdot 90 \quad ?$

$6750 = 6750 \qquad \text{True.}$

It is a true proportion.

16. $\dfrac{\frac{1}{3}}{\frac{2}{7}} = \dfrac{\frac{20}{4}}{\frac{9}{2}}$

$\dfrac{1}{3} \cdot \dfrac{9}{2} = \dfrac{2}{7} \cdot \dfrac{20}{4} \quad ?$

$\dfrac{3}{2} = \dfrac{10}{7} \qquad \text{False}$

It is not a true proportion.

17. $\dfrac{x}{4} = \dfrac{35}{20}$

$x \cdot 20 = 4 \cdot 35$

$20x = 140$

$\dfrac{20x}{20} = \dfrac{140}{20}$

$x = 7$

18. $\dfrac{\frac{18}{10}}{\frac{7}{11}} = \dfrac{x}{\frac{5}{9}}$

$\dfrac{18}{10} \cdot \dfrac{5}{9} = \dfrac{7}{11} \cdot x$

$1 = \dfrac{7}{11} \cdot x$

$\dfrac{11}{7} \cdot 1 = \dfrac{11}{7} \cdot \dfrac{7}{11}x$

$\dfrac{11}{7} = x$

19. $\dfrac{3.6}{4} = \dfrac{9}{y}$

$3.6 \cdot y = 4 \cdot 9$
$3.6y = 36$

$\dfrac{3.6y}{3.6} = \dfrac{36}{3.6}$

$y = 10$

20. Let x = number of miles

$\dfrac{1 \text{ inch}}{50 \text{ miles}} = \dfrac{17 \text{ inches}}{x \text{ miles}}$

$1 \cdot x = 50 \cdot 17$
$x = 850$

They are 850 miles apart.

21. Let x = number of inches

$\dfrac{1 \text{ inch}}{50 \text{ miles}} = \dfrac{x \text{ inches}}{45 \text{ miles}}$

$1 \cdot 45 = 50 \cdot x$
$45 = 50x$

$\dfrac{45}{50} = \dfrac{50x}{50}$

$\dfrac{9}{10} = x$

On the map they will be $\dfrac{9}{10}$ of an inch apart.

22. $\dfrac{10}{6} = \dfrac{14}{x}$

$10 \cdot x = 6 \cdot 14$
$10x = 84$

$\dfrac{10x}{10} = \dfrac{84}{10}$

$x = 8.4$

23. $\dfrac{6.4 \text{ ft}}{1 \text{ tablecloth}} = \dfrac{x \text{ ft}}{5 \text{ tablecloths}}$

$6.4 \cdot 5 = 1 \cdot x$
$32 = x$

32 ft of material will be used.

24. $\dfrac{\text{height}}{\text{shadow}} : \dfrac{5\frac{1}{4} \text{ ft}}{3\frac{1}{2} \text{ ft}} = \dfrac{x \text{ ft}}{32 \text{ ft}}$

$5\dfrac{1}{4} \cdot 32 = 3\dfrac{1}{2} \cdot x$

$\dfrac{21}{4} \cdot 32 = \dfrac{7}{2}x$

$168 = \dfrac{7}{2}x$

$\dfrac{2}{7} \cdot 168 = \dfrac{2}{7} \cdot \dfrac{7}{2}x$

$48 = x$

The tree is 48 feet tall.

7.1 SOLUTIONS TO EXERCISES

1. $52\% = 52\% = 0.52$
   ↑

2. $5\% = 05\% = 0.05$
   ↑

3. $81.6\% = 81.6\% = 0.816$
   ↑

4. $190\% = 190\% = 1.9$
   ↑

5. $0.99 = 0.99 = 99\%$
   ↑

6. $0.037 = 0.037 = 3.7\%$
   ↑

7. $2.12 = 2.12 = 212\%$
   ↑

8. $14 = 14.00 = 1400\%$
   ↑

9. $16\% = \dfrac{16}{100}$

   $= \dfrac{4}{25}$

10. $118\% = \dfrac{118}{100}$

    $= \dfrac{59}{50}$

11. $3.2\% = \dfrac{3.2}{100}$

    $= \dfrac{3.2}{100} \cdot \dfrac{10}{10}$

    $= \dfrac{32}{1000}$

    $= \dfrac{4}{125}$

12. $0.6\% = \dfrac{0.6}{100}$

    $= \dfrac{0.6}{100} \cdot \dfrac{10}{10}$

    $= \dfrac{6}{1000}$

    $= \dfrac{3}{500}$

13. $\dfrac{3}{8} = \dfrac{3}{8} \cdot 100\%$

    $= \dfrac{300}{8}\%$

    $= 37.5\%$

14. $\dfrac{7}{20} = \dfrac{7}{20} \cdot 100\%$

    $= \dfrac{700}{20}\%$

    $= 35\%$

15. $1\dfrac{9}{10} = \dfrac{19}{10}$

    $= \dfrac{19}{10} \cdot 100\%$

    $= 190\%$

16. $\dfrac{9}{40} = \dfrac{9}{40} \cdot 100\%$

    $= \dfrac{900}{40}\%$

    $= 22.5\%$

17. 100% of 25 is all of 25.
    Hence 25 students attended class.

18. $0.05 = 0.05$
    ↑
    $= 5\%$

19. $3.5\% = 03.5$
$$\underset{\uparrow}{}$$
$$ = 0.035$$

20. $\dfrac{3}{4} = \dfrac{3}{4} \cdot 100\%$

$$\phantom{\dfrac{3}{4}} = \dfrac{300}{4}\%$$

$$\phantom{\dfrac{3}{4}} = 75\%$$

7.2 SOLUTIONS TO EXERCISES

1. 35%   of   80   is   what number?
 35% $\cdot$ 80   = x

2. What percent   of   12   is   8?
 x $\cdot$ 12   =   8

3. 6.2   is   29%   of   what number?
 6.2   =   29% $\cdot$ x

4. 102%   of   50   is   what?
 102% $\cdot$ 50   = x

5. $5\% \cdot 40 = x$
 $0.05 \cdot 40 = x$
 $2 = x$

 2 is 5% of 40.

6. $x = 18\% \cdot 70$
 $x = 0.18 \cdot 70$
 $x = 12.6$

 12.6 is 18% of 70.

7. $40 = 20\% \cdot x$
 $40 = 0.20x$

$$\frac{40}{0.20} = \frac{0.20x}{0.20}$$

 $200 = x$

 40 is 20% of 200.

8. $0.36 = 52\% \cdot x$
 $0.36 = 0.52x$

$$\frac{0.36}{0.52} = \frac{0.52x}{0.52}$$

 $0.69 = x$

 0.36 is 52% of 0.69.

9. $9 = x \cdot 36$
 $9 = 36x$

$$\frac{9}{36} = \frac{36x}{36}$$

 $0.25 = x$
 $25\% = x$

 9 is 25% of 36.

10. $8.25 = x \cdot 82.5$
 $8.25 = 82.5x$

$$\frac{8.25}{82.5} = \frac{82.5x}{82.5}$$

 $0.1 = x$
 $10\% = x$

 8.25 is 10% of 82.5.

11. $0.8 = 20\% \cdot x$
 $0.8 = 0.20x$

$$\frac{0.8}{0.20} = \frac{0.20x}{0.20}$$

 $4 = x$

 0.8 is 20% of 4.

12. $70 = 100\% \cdot x$
 $70 = 1 \cdot x$
 $70 = x$

 70 is 100% of 70.

13. $14.2 = 8\frac{1}{4}\% \cdot x$

$14.2 = 0.0825x$

$\dfrac{14.2}{0.0825} = \dfrac{0.0825x}{0.0825}$

$172.12 = x$

14.2 is $8\frac{1}{4}\%$ of 172.12.

14. $520 = x \cdot 65$
$520 = 65x$

$\dfrac{520}{65} = \dfrac{65x}{65}$

$8 = x$
$800\% = x$

520 is 800% of 65.

15. $21.3 = x \cdot 100$
$21.3 = 100x$

$\dfrac{21.3}{100} = \dfrac{100x}{100}$

$0.213 = x$
$21.3\% = x$

21.3 is 21.3% of 100.

16. $90\% \cdot x = 525.6$
$0.90x = 525.6$

$\dfrac{0.90x}{0.90} = \dfrac{525.6}{0.90}$

$x = 584$

90% of 584 is 525.6.

7.3 SOLUTIONS TO EXERCISES

1.   What percent of 51 is 17?
         ↓   ↓   ↓
    percent   base   amount

$\dfrac{17}{51} = \dfrac{p}{100}$

2.   9% of what number is 10?
     ↓   ↓   ↓
  percent   base   amount

$\dfrac{9}{100} = \dfrac{10}{b}$

3.   11 is   37% of what number?
   ↓   ↓   ↓
  amount   percent   base

$\dfrac{11}{b} = \dfrac{37}{100}$

4.   70% of 112 is what number?
    ↓   ↓   ↓
  percent   base   amount

$\dfrac{70}{100} = \dfrac{a}{112}$

5. $\dfrac{15}{100} = \dfrac{a}{70}$

$15 \cdot 70 = 100 \cdot a$
$1050 = 100a$

$\dfrac{1050}{100} = \dfrac{100a}{100}$

$10.5 = a$

10.5 is 15% of 70.

6. $\dfrac{12}{100} = \dfrac{a}{50}$

$12 \cdot 50 = 100 \cdot a$
$600 = 100a$

$\dfrac{600}{100} = \dfrac{100a}{100}$

$6 = a$

6 is 12% of 50.

7. $$\frac{60}{b} = \frac{8}{100}$$

$$60 \cdot 100 = b \cdot 8$$
$$6000 = 8b$$

$$\frac{6000}{8} = \frac{8b}{8}$$

$$750 = b$$

60 is 8% of 750.

8. $$\frac{0.7}{b} = \frac{80}{100}$$

$$0.7 \cdot 100 = b \cdot 80$$
$$70 = 80b$$

$$\frac{70}{80} = \frac{80b}{80}$$

$$0.88 = b$$

0.7 is 80% of 0.88.

9. $$\frac{60}{110} = \frac{p}{100}$$

$$60 \cdot 100 = 110 \cdot p$$
$$6000 = 110p$$

$$\frac{6000}{110} = \frac{110p}{110}$$

$$54.55 = p$$

60 is 54.55% of 110.

10. $$\frac{12}{112} = \frac{p}{100}$$

$$12 \cdot 100 = 112 \cdot p$$
$$1200 = 112p$$

$$\frac{1200}{112} = \frac{112p}{112}$$

$$10.71 = p$$

12 is 10.71% of 112.

11. $$\frac{120}{b} = \frac{95}{100}$$

$$120 \cdot 100 = b \cdot 95$$
$$12000 = 95b$$

$$\frac{12000}{95} = \frac{95b}{95}$$

$$126.32 = b$$

120 is 95% of 126.32.

12. $$\frac{45}{b} = \frac{100}{100}$$

$$45 \cdot 100 = b \cdot 100$$
$$4500 = 100b$$

$$\frac{4500}{100} = \frac{100b}{100}$$

$$45 = b$$

45 is 100% of 45.

13. $$\frac{31.25}{b} = \frac{6\frac{1}{4}}{100}$$

$$\frac{31.25}{b} = \frac{6.25}{100}$$

$$31.25 \cdot 100 = b \cdot 6.25$$
$$3125 = 6.25b$$

$$\frac{3125}{6.25} = \frac{6.25b}{6.25}$$

$$500 = b$$

31.25 is $6\frac{1}{4}$% of 500.

14. $$\frac{200}{50} = \frac{p}{100}$$

$$200 \cdot 100 = 50 \cdot p$$
$$20000 = 50p$$

$$\frac{20000}{50} = \frac{50p}{50}$$

$$400 = p$$
200 is 400% of 50.

15. $$\frac{27.6}{100} = \frac{p}{100}$$

$$27.6 \cdot 100 = 100 \cdot p$$
$$2760 = 100p$$

$$\frac{2760}{100} = \frac{100p}{100}$$

$$27.6 = p$$

27.6 is 27.6% of 100.

16. $$\frac{110}{100} = \frac{60}{b}$$

$$110 \cdot b = 100 \cdot 60$$
$$110b = 6000$$

$$\frac{110b}{110} = \frac{6000}{110}$$

$$b = 54.55$$

110% of 54.55 is 60.

7.4 SOLUTIONS TO EXERCISES

1. $4.5\% \cdot 1600 = x$
   $0.045 \cdot 1600 = x$
   $72 = x$

   The sales tax is $72.

2. 325 is what percent of 500?

   $325 = x \cdot 500$
   $325 = 500x$

   $$\frac{325}{500} = \frac{500x}{500}$$

   $0.65 = x$
   $65\% = x$

   65% of the money was spent on textbooks.

3. 15.02% of $1250 is what?

   $15.02\% \cdot 1250 = x$
   $0.1502 \cdot 1250 = x$
   $187.75 = x$

   The social security tax is $187.75.

4. $7.38 is what percent of $90?

   $7.38 = x \cdot 90$
   $7.38 = 90x$

   $$\frac{7.38}{90} = \frac{90x}{90}$$

   $0.082 = x$
   $8.2\% = x$

   The dividend is 8.2% of the stock price.

5. amount of increase = 26.2 − 22.4 = 3.8

   $$\text{percent of increase} = \frac{\text{amount of increase}}{\text{original amount}}$$

   $$= \frac{3.8}{22.4} \approx 0.170 = 17.0\%$$

6. amount of increase = 7110 − 6230 = 880

   $$\text{percent of increase} = \frac{\text{amount of increase}}{\text{original amount}}$$

   $$= \frac{880}{6230} \approx 0.141 = 14.1\%$$

7. amount of decrease = 2100 − 1600 = 500

   $$\text{percent of decrease} = \frac{\text{amount of decrease}}{\text{original amount}}$$

   $$= \frac{500}{2100} \approx 0.238 = 23.8\%$$

8. amount of decrease = 208,000 − 178,000 = 30,000

   $$\text{percent of decrease} = \frac{\text{amount of decrease}}{\text{original amount}}$$

   $$= \frac{30,000}{208,000} \approx 0.144 = 14.4\%$$

9. amount of increase = 1.32 − 1.16 = 0.16

   $$\text{percent of increase} = \frac{\text{amount of increase}}{\text{original amount}}$$

   $$= \frac{0.16}{1.16} \approx 0.138 = 13.8\%$$

10. amount of decrease = 42 − 36 = 6

$$\text{percent of decrease} = \frac{\text{amount of decrease}}{\text{original amount}}$$

$$= \frac{6}{42} \approx 0.143 = 14.3\%$$

7.5 SOLUTIONS TO EXERCISES

1. Sales tax = tax rate · purchase price
    = 8% · $56
    = 0.08($56)
    = $4.48

2. Sales tax = tax rate · purchase price
    = 5% · $250
    = 0.05($250)
    = $12.50

3. Sales tax = tax rate · purchase price
    = 4.5% · $350
    = 0.045($350)
    = $15.75

4. Sales tax = tax rate · purchase price
    = 7.5% · $1200
    = 0.075($1200)
    = $90

    Total price = purchase price + sales tax
    = $1200 + $90
    = $1290

5. Sales tax = tax rate · purchase price
    = 6% · $799
    = 0.06($799)
    = $47.94

    Total price = purchase price + sales tax
    = $799 + $47.94
    = $846.94

6. purchase price = $45 + $75 + $120
    = $240

    sales tax = tax rate · purchase price
    = 5.5% · $240
    = 0.055($240)
    = $13.20

    Total price = purchase price + sales tax
    = $240 + $13.20
    = $253.20

7. commission = commission · price of
                   rate   house

    = 1.2% · $245,000
    = 0.012($245,000)
    = $2940

8. commission = commission rate · sales
       2675   = r · 29,722

    2675 = 29,722r

    $$\frac{2675}{29{,}722} = \frac{29{,}722r}{29{,}722}$$

    0.090 = r
    r = 9.0%

    The commission rate is 9.0%.

9. commission = commission rate · sales
       c =   5.5%   · 18,650

    c = 0.055(18,650)
    c = 1025.75

    Her commission was $1025.75.

10. discount = 20% · 800
     = 0.20(800)
     = 160

     The decrease in price is $160.

     new price = original price − discount
     = 800 − 160
     = 640

     The new reduced price is $640.

11. discount = 12% · 125
     = 0.12(125)
     = 15

     The decrease in price is $15.

     new price = original price − discount
     = 125 − 15
     = 110

     The new reduced price is $110.

12. discount = 3% · 25,000
$$= 0.03(25,000)$$
$$= 750$$

The decrease in price is $750.

new price = original price − discount
$$= 25,000 − 750$$
$$= 24,250$$

The new reduced price is $24,250.

7.6 SOLUTIONS TO EXERCISES

1. $I = P \cdot R \cdot T$
   $$= \$750 \cdot 4\% \cdot 5$$
   $$= \$750 \cdot 0.04 \cdot 5$$
   $$= \$150$$

2. $I = P \cdot R \cdot T$
   $$= \$8000 \cdot 5.5\% \cdot 6$$
   $$= \$8000 \cdot 0.055 \cdot 6$$
   $$= \$2640$$

3. $I = P \cdot R \cdot T$
   $$= \$400 \cdot 14\% \cdot 2\frac{1}{2}$$
   $$= \$400 \cdot 0.14 \cdot 2.5$$
   $$= \$140$$

4. $I = P \cdot R \cdot T$

   20 months $= \dfrac{20}{12}$ years

   $$= 1\frac{2}{3} \text{ years}$$
   $$= \$650 \cdot 16.5\% \cdot 1\frac{2}{3}$$
   $$= \$650 \cdot 0.165 \cdot \frac{5}{3}$$
   $$= \$178.75$$

5. $I = P \cdot R \cdot T$

   3 months $= \dfrac{3}{12}$ year

   $$= 0.25 \text{ year}$$
   $$= \$200 \cdot 13\% \cdot 0.25$$
   $$= \$200 \cdot 0.13 \cdot 0.25$$
   $$= \$6.50$$

6. $I = P \cdot R \cdot T$
   $$= \$3000 \cdot 7\% \cdot 6\frac{1}{4}$$
   $$= \$3000 \cdot 0.07 \cdot 6.25$$
   $$= \$1312.50$$

7. $I = P \cdot R \cdot T$
   $$= \$75,000 \cdot 11.5\% \cdot 3$$
   $$= \$75,000 \cdot 0.115 \cdot 3$$
   $$= \$25,875$$

8. $I = P \cdot R \cdot T$
   $$= \$6250 \cdot 8.5\% \cdot 1\frac{1}{2}$$
   $$= \$6250 \cdot 0.085 \cdot 1.5$$
   $$= \$796.88$$

9. annually
   rate = 12%
   time = 10 years
   compound interest factor = 3.10585
   (From Appendix F in textbook)

   Total amount = original principal · compound interest factor
   $$= \$9500(3.10585)$$
   $$= \$29,505.58$$

10. semiannually
    rate = 17%
    time = 5 years
    compound interest factor = 2.26098
    (From Appendix F in textbook)

    Total amount = original principal · compound interest factor
    $$= \$3025(2.26098)$$
    $$= \$6839.46$$

11. quarterly
     rate = 8%
     time = 15 years

     compound interest factor = 3.28103
     (From Appendix F in textbook)

     Total amount = original principal · compound interest factor
                  = $12,000(3.28103)
                  = $39,372.36

12. daily
     rate = 6%
     time = 20 years

     compound interest factor = 3.31979
     (From Appendix F in textbook)

     Total amount = original principal · compound interest factor
                  = $22,000(3.31979)
                  = $73,035.38

13. annually
     rate = 5%
     time = 5 years

     compound interest factor = 1.27628
     (From Appendix F in textbook)

     Total amount = original principal · compound interest factor
                  = $950(1.27628)
                  = $1212.47

14. quarterly
     rate = 16%
     time = 10 years

     compound interest factor = 4.80102
     (From Appendix F in textbook)

     Total amount = original principal · compound interest factor
                  = $1475(4.80102)
                  = $7081.50

15. Total amount = amount borrowed + interest
                  = $30,000 + $14,578.42
                  = $44,578.42

Number of payments = 5 years · 12 payments/year
= 60 payments

$$\text{Monthly payment} = \frac{\text{Total amount}}{\text{number of payments}}$$

$$= \frac{\$44,578.42}{60}$$

$$= \$742.97$$

16. Total amount = amount borrowed + interest
= $128,000 + $232,000
= $360,000

Number of payments = 30 years · 12 payments/year
= 360 payments

$$\text{Monthly payment} = \frac{\text{Total amount}}{\text{number of payments}}$$

$$= \frac{\$360,000}{360}$$

$$= \$1000$$

CHAPTER 7 PRACTICE TEST SOLUTIONS

1. 0.12 7 = 12.7%
   $\underline{\quad\uparrow}$

2. 0.3% = 00.3%
   $\uparrow$
   = 0.003

3. $140\% = \dfrac{140}{100}$

   $= \dfrac{7}{5}$

4. $\dfrac{19}{20} = \dfrac{19}{20} \cdot 100\%$

   $= \dfrac{1900}{20}\%$

   $= 95\%$

5. $x = 35\% \cdot 90$
   $x = 0.35 \cdot 90$
   $x = 31.5$

6. $0.4\% \cdot x = 10.8$
   $0.004x = 10.8$

   $\dfrac{0.004x}{0.004} = \dfrac{10.8}{0.004}$

   $x = 2700$

7. $507 = x \cdot 845$

   $\dfrac{507}{845} = \dfrac{x \cdot 845}{845}$

   $0.6 = x$
   $60\% = x$

8. 39.2   is   18%   of   what number?

   39.2   =   0.18   · x

9. $10.4\% = \dfrac{10.4}{100}$

   $= \dfrac{10.4}{100} \cdot \dfrac{10}{10}$

   $= \dfrac{104}{1000}$

   $= \dfrac{13}{125}$

10. $\dfrac{\text{amount}}{\text{base}} = \dfrac{\text{percent}}{100}$

    $\dfrac{29}{185} = \dfrac{p}{100}$

11. Let x = pounds of copper

    $x = 16\% \cdot 180$
    $= 0.16 \cdot 180$
    $= 28.8$

    It contains 28.8 lb of copper.

12. $14,575 is 25% of what total value?

$$14575 = 0.25 \cdot x$$

$$\frac{14575}{0.25} = \frac{0.25x}{0.25}$$

$$58300 = x$$

The total value is $58,300.

13. Let x = commission

$$x = 1.6\% \cdot 185,000$$
$$x = 0.016 \cdot 185,000$$
$$x = 2960$$

The commission was $2960.

14. Total amount = $650 + 4.5\% \cdot 650$
$$= 650 + 0.045 \cdot 650$$
$$= 650 + 29.25$$
$$= 679.25$$

The total amount charged was $679.25.

15. 9 months = $\frac{9}{12}$ yr

$$= \frac{3}{4} \text{ yr}$$

$$I = P \cdot R \cdot T$$

$$= \$600 \cdot 15.5\% \cdot \frac{3}{4}$$

$$= \$600 \cdot 0.155 \cdot \frac{3}{4}$$

$$= \$69.75$$

The simple interest was $69.75.

16. $I = P \cdot R \cdot T$

$$= \$4000 \cdot 7.75\% \cdot 5\frac{1}{2}$$

$$= \$4000 \cdot 0.0775 \cdot 5.5$$
$$= \$1705$$

17.   increase = $16\% \cdot \$2.15$
$$= 0.16 \cdot \$2.15$$
$$= \$0.344 \approx \$0.34$$

new price = old price + increase
$$= \$2.15 + \$0.34$$
$$= \$2.49$$

18. amount of decrease = $580 - 510 = 70$

percent of decrease = $\dfrac{\text{amount of decrease}}{\text{original amount}}$

$$= \frac{70}{580} \approx 0.121 = 12.1\%$$

19. quarterly
    rate = 8%
    time = 20 years

compound interest factor = 4.87544
(From Appendix F in textbook)

Total amount = original principal $\cdot$ compound interest factor
$$= \$5 \,(4.87544)$$
$$= \$24.38.$$

20. semiannually
    rate = 6%
    time = 10 years

compound interest factor = 1.80611
(From Appendix F in textbook)

Total amount = original principal $\cdot$ compound interest factor
$$= \$5260 \,(1.80611)$$
$$= \$9500.14$$

21. amount of increase = $3.09 - 2.50 = 0.59$

percent of increase = $\dfrac{\text{amount of increase}}{\text{original amount}}$

$$= \frac{0.59}{2.50} = 0.236$$

$$= 23.6\%$$

22. Total borrowed = $8800 + $620.35
    = $9420.35

$2\frac{1}{2}$ years = 30 months

monthly payment = $\dfrac{\$9420.35}{30}$

= $314.01

23. commission = commission rate · sales

= 1.8% · 200,000
= (0.018)(200,000)
= 3600

Her commission was $3600.

8.1 SOLUTIONS TO EXERCISES

1. amount = percent · base
   amount =   8%   · 500
           = (0.08)(500)
           =   40

   40 students prefer horror movies.

2. sci-fi:
          amount =   percent · base
          amount =   10% · 500
                = (0.10)(500)
                =   50

   action:
          amount =   37% · 500
                = (0.37)(500)
                =   185

   185 + 50 = 235

   235 students prefer either sci-fi
   or action movies.

3. percent of students preferring action
   movies = 37%

   $$\text{ratio} = \frac{37}{100}$$

4. $$\frac{\text{comedy \%}}{\text{drama \%}} = \frac{20\%}{25\%}$$
   $$= \frac{20}{25}$$
   $$= \frac{4}{5}$$

5. Look for the largest sector. Swimming
   had the most student participation.

6. Look for the smallest sector. Canoeing
   had the least student participation.

7. $$\frac{\text{number of students fishing}}{\text{total number of students}} = \frac{28}{150} = \frac{14}{75}$$

8. $$\frac{\text{number of students canoeing}}{\text{total number of students playing volleyball}} = \frac{16}{18}$$
   $$= \frac{8}{9}$$

9. $$\frac{\text{number of students swimming or playing softball}}{\text{total number of students}}$$
   $$= \frac{55 + 35}{150}$$
   $$= \frac{88}{150}$$
   $$= \frac{44}{75}$$

10. The second largest category is softball.

11.

Sector	Degrees in Each Sector
Fenton	40% × 360° = 144°
Smith	6% × 360° = 21.6°
Mosser	12% × 360° = 43.2°
Victorian	18% × 360° = 64.8°
Westmoreland	24% × 360° = 86.4°

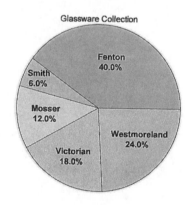

Glassware Collection

8.2 SOLUTIONS TO EXERCISES

1. (number of houses) · (5 houses) = 4 · 5 houses
                                 = 20 houses

2. (number of houses) $\cdot$ (5 houses) = 5 $\cdot$ 5 houses
                                  = 25 houses

3. 1999 (this line contains the most symbols)

4. 1998: 5 $\cdot$ 5 houses = 25 houses
   1994: 3 $\cdot$ 5 houses = 15 houses

   25 houses − 15 houses = 10 houses

   There were 10 more houses renovated in 1998 than in 1994.

5. First add up the total number of symbols:
   3 + 4 + 4 + 2 + 5 + 6 = 24

   total number of renovations = 24 $\cdot$ 5 houses
                                  = 120 houses

6. 20 renovations = $\dfrac{20}{5}$ = 4 symbols

   1995 and 1996 were years with 20 renovations.

7. Find the bar corresponding to the year 1997, go to the top of the bar and then across to the vertical axis to identify the profit. It appears to be halfway between $350,000 and $400,000. Hence the profit during 1997 was $375,000.

8. Find $450,000 on the vertical axis, then go horizontally to see if this corresponds to the top of a bar. There was a profit of $450,000 in the year 1998.

9. 1998: $450,000
   1996: − $350,000
            $100,000

   There was $100,000 more in profits in 1998 than in 1996.

10. Look at the top of consecutive bars, left to right and see where there is the greatest gap. The greatest increase in profit occurred between 1997 and 1998.

11. 1996: $350,000
    1997: $375,000
    1998: $450,000
    1999: $500,000
    Total: $1,675,000

12. The dot on the line graph corresponding to Tuesday appears to be halfway between 76 and 78. Hence the high temperature reading for Tuesday was 77°F.

13. Look for the lowest dot on the graph. The temperature was lowest on Monday.

14. Looking at consecutive days from left to right, locate the longest line segment that is rising. The greatest increase occurred between Tuesday and Wednesday.

15. Find 82 on the vertical axis, then go horizontally across to the line graph to see if this temperature corresponds to a day. 82°F was the high on Wednesday.

16. Sun: 80°F
    Sat: 78°F
    difference: 2°F

    There was a 2° difference in temperature readings on Sunday and Saturday.

8.3 SOLUTIONS TO EXERCISES

1. − 6.

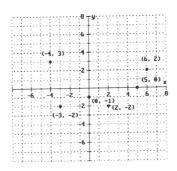

7. $A(-3, 3)$; $B(4, 5)$; $C(-1, -1)$; $D(0, 6)$

8. $y = 8x$
   $8 = 8(-1)$
   $8 = -8$
   False
   $(-1, 8)$ is not a solution.

9. $x = -4y$
   $0 = -4(0)$
   $0 = 0$
   True
   $(0, 0)$ is a solution.

10. $x + 3y = 5$
   $1 + 3(2) = 5$
   $1 + 6 = 5$
     $7 = 5$
     False
   $(1, 2)$ is not a solution.

11. $3x - 5y = 8$
   $3(1) - 5(-1) = 8$
     $3 + 5 = 8$
      $8 = 8$
       True
   $(1, -1)$ is a solution.

12.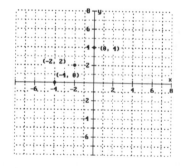

13.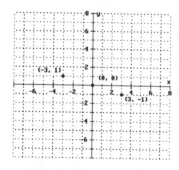

14. $x = 10y$
   $10 = 10y$

   $\dfrac{10}{10} = \dfrac{10y}{10}$

   $1 = y$
   $(10, 1)$

   $x = 10y$
   $x = 10(0)$
   $x = 0$
   $(0, 0)$

   $x = 10y$
   $x = 10(2)$
   $x = 20$
   $(20, 2)$

15. $4x - y = 8$
     $4(2) - y = 8$
      $8 - y = 8$
  $8 - y + (-8) = 8 + (-8)$
        $-y = 0$

     $\dfrac{-y}{-1} = \dfrac{0}{-1}$

      $y = 0$
     $(2, 0)$

     $4x - y = 8$
     $4(3) - y = 8$
     $12 - y = 8$
  $12 - y + (-12) = 8 + (-12)$
       $-y = -4$

     $\dfrac{-y}{-1} = \dfrac{-4}{-1}$

      $y = 4$
     $(3, 4)$

     $4x - y = 8$
     $4x - 8 = 8$
  $4x - 8 + 8 = 8 + 8$
      $4x = 16$

     $\dfrac{4x}{4} = \dfrac{16}{4}$

      $x = 4$
     $(4, 8)$

16. $x + 6y = 0$
     $x + 6(1) = 0$
     $x + 6 = 0$
     $x + 6 + (-6) = 0 + (-6)$
     $x = -6$
     $(-6, 1)$

     $x + 6y = 0$
     $x + 6(-2) = 0$
     $x - 12 = 0$
     $x - 12 + 12 = 0 + 12$
     $x = 12$
     $(12, -2)$

     $x + 6y = 0$
     $0 + 6y = 0$
     $6y = 0$

     $$\frac{6y}{6} = \frac{0}{6}$$

     $y = 0$
     $(0, 0)$

2. $x + 2y = 6$

    Let $x = 0$: $0 + 2y = 6$
    $2y = 6$

    $$\frac{2y}{2} = \frac{6}{2}$$

    $y = 3$
    $(0, 3)$

    Let $y = 0$: $x + 2(0) = 6$
    $x + 0 = 6$
    $x = 6$
    $(6, 0)$

    Let $x = 2$: $2 + 2y = 6$
    $2 + 2y + (-2) = 6 + (-2)$
    $2y = 4$

    $$\frac{2y}{y} = \frac{4}{2}$$

    $y = 2$
    $(2, 2)$

8.4 SOLUTIONS TO EXERCISES

1. $x + y = -5$

    Let $x = 0$: $0 + y = -5$
    $y = -5$
    $(0, -5)$

    Let $y = 0$: $x + 0 = -5$
    $x = -5$
    $(-5, 0)$

    Let $x = -1$: $-1 + y = -5$
    $-1 + y + 1 = -5 + 1$
    $y = -4$
    $(-1, -4)$

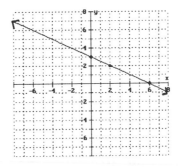

3. $3x - y = 6$

    Let $x = 0$: $3(0) - y = 6$
    $0 - y = 6$
    $-y = 6$

    $$\frac{-y}{-1} = \frac{6}{-1}$$

    $y = -6$
    $(0, -6)$

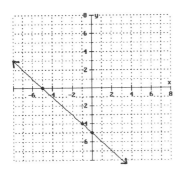

Let $y = 0$: $3x - 0 = 6$

$$3x = 6$$

$$\frac{3x}{3} = \frac{6}{3}$$

$$x = 2$$
$$(2, 0)$$

Let $x = 1$: $3(1) - y = 6$

$$3 - y = 6$$
$$3 - y + (-3) = 6 + (-3)$$
$$-y = 3$$

$$\frac{-y}{-1} = \frac{3}{-1}$$

$$y = -3$$
$$(1, -3)$$

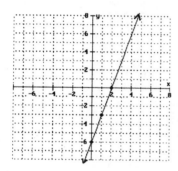

4. $y - x = -1$

Let $x = 0$: $y - 0 = -1$
$$y = -1$$
$$(0, -1)$$

Let $y = 0$: $0 - x = -1$
$$-x = -1$$

$$\frac{-x}{-1} = \frac{-1}{-1}$$

$$x = 1$$
$$(1, 0)$$

Let $x = 2$: $y - 2 = -1$
$$y - 2 + 2 = -1 + 2$$
$$y = 1$$
$$(2, 1)$$

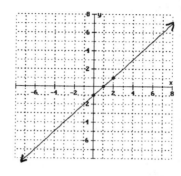

5. $y = 2x - 4$

x	$y = 2x - 4$
0	$2(0) - 4 = -4$
1	$2(1) - 4 = -2$
2	$2(2) - 4 = 0$

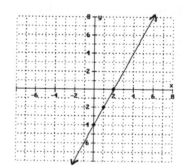

6. $x = 3y + 1$

Let $x = 4$: $4 = 3y + 1$
$$4 + (-1) = 3y + 1 + (-1)$$
$$3 = 3y$$

$$\frac{3}{3} = \frac{3y}{3}$$

$$1 = y$$
$$(4, 1)$$

Let $y = 0$: $x = 3(0) + 1$
$$x = 1$$
$$(1, 0)$$

Let $y = -1$: $x = 3(-1) + 1$
$$x = -3 + 1$$
$$x = -2$$
$$(-2, -1)$$

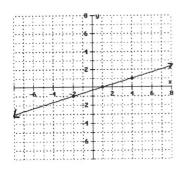

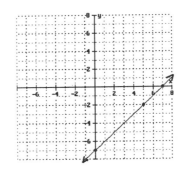

7. $y = x + 3$

x	$y = x + 3$
-1	$-1 + 3 = 2$
0	$0 + 3 = 3$
1	$1 + 3 = 4$

9. $x = 6$ is a vertical line that crosses the x-axis at 6.

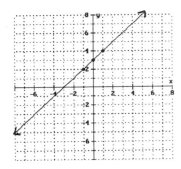

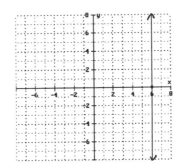

10. $y = -4$ is a horizontal line that crosses the y-axis at -4.

8. $x = y + 7$

Let $x = 0$: $0 = y + 7$
$$0 + (-7) = y + 7 + (-7)$$
$$-7 = y$$
$$(0, -7)$$

Let $y = 0$: $x = 0 + 7$
$$x = 7$$
$$(7, 0)$$

Let $y = -2$: $x = -2 + 7$
$$x = 5$$
$$(5, -2)$$

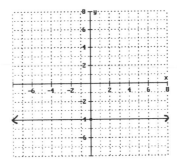

11. $y = 7$ is a horizontal line that crosses the y-axis at 7.

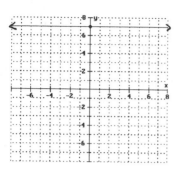

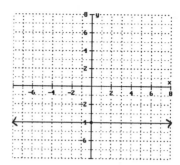

12. $x = -5$ is a vertical line that crosses the x-axis at -5.

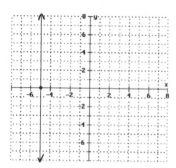

15. $x - 4y = 0$

Let $x = 0$:
$$0 - 4y = 0$$
$$-4y = 0$$

$$\frac{-4y}{-4} = \frac{0}{-4}$$

$$y = 0$$
$$(0, 0)$$

Let $y = 1$:
$$x - 4(1) = 0$$
$$x - 4 = 0$$
$$x - 4 + 4 = 0 + 4$$
$$x = 4$$
$$(4, 1)$$

Let $y = -1$:
$$x - 4(-1) = 0$$
$$x + 4 = 0$$
$$x + 4 + (-4) = 0 + (-4)$$
$$x = -4$$
$$(-4, -1)$$

13. $x - 2 = 0$
$x - 2 + 2 = 0 + 2$
$x = 2$

This is a vertical line that crosses the x-axis at 2.

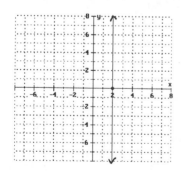

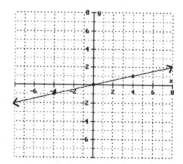

14. $y + 4 = 0$
$y + 4 + (-4) = 0 + (-4)$
$y = -4$

This is a horizontal line that crosses the y-axis at -4.

16. $y + 3x = 0$

Let $x = 0$:
$$y + 3(0) = 0$$
$$y + 0 = 0$$
$$y = 0$$
$$(0, 0)$$

Let $x = 1$:
$$y + 3(1) = 0$$
$$y + 3 = 0$$
$$y + 3 + (-3) = 0 + (-3)$$
$$y = -3$$
$$(1, -3)$$

Let $x = -1$:
$$y + 3(-1) = 0$$
$$y - 3 = 0$$
$$y - 3 + 3 = 0 + 3$$
$$y = 3$$
$$(-1, 3)$$

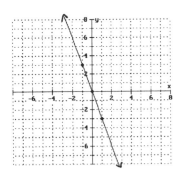

8.5 SOLUTIONS TO EXERCISES

1. mean $= \dfrac{18 + 24 + 16 + 31 + 19 + 26 + 30}{7}$

           $= \dfrac{164}{7}$

           $= 23.4$

2. mean $= \dfrac{51 + 40 + 80 + 71 + 62 + 54 + 95 + 48}{8}$

           $= \dfrac{501}{8}$

           $= 62.625$

3. mean $= \dfrac{9.4 + 8.6 + 11.2 + 17.8 + 10.1 + 3.4}{6}$

           $= \dfrac{60.5}{6}$

           $= 10.08$

4. mean $= \dfrac{122 + 146 + 130 + 124 + 148 + 132 + 120 + 160 + 154 + 155}{10}$

           $= \dfrac{1391}{10}$

           $= 139.1$

5. 0.1, 0.2, <u>0.3</u>, <u>0.6</u>, 0.7, 0.9

   median $= \dfrac{0.3 + 0.6}{2}$

   $= \dfrac{0.9}{2}$

   $= 0.45$

6. 9, 11, 27, 54, <u>62</u>, 65, 71, 81, 90

   median $= 62$

7. 328, 387, 419, <u>491</u>, 505, 576, 637

   median $= 491$

8. 45, 60, 75, <u>80</u>, <u>87</u>, 90, 91, 105

   median $= \dfrac{80 + 87}{2}$

   $= \dfrac{167}{2}$

   $= 83.5$

9. mode $= 15$ (it appears the most number of times)

10. mode $= 140$ (it appears the most number of times)

11. modes $= 5.4$ and 4.5

12. modes $= 22$ and 14

13.

Grade	Point Value of Grade	Credit Hours	(Point value) · (Credit Hours)
A	4	4	16
C	2	3	6
B	3	3	9
B	3	4	12
	Totals	14	43

grade point average $= \dfrac{43}{14} = 3.07$

14.

Grade	Point Value of Grade	Credit Hours	(Point value) · (Credit Hours)
C	2	3	6
C	2	5	10
A	4	3	12
A	4	4	16
B	3	3	9
	Totals	18	53

grade point average $= \dfrac{53}{18} = 2.94$

15.

Grade	Point Value of Grade	Credit Hours	(Point value) · (Credit Hours)
B	3	3	9
D	1	3	3
C	2	5	10
D	1	4	4
F	0	3	0
Totals		18	26

$$\text{grade point average} = \frac{26}{18} = 1.44$$

8.6 SOLUTIONS TO EXERCISES

1. Choose a   Toss a   Outcomes
    number   coin

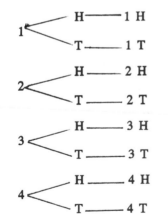

There are 8 possible outcomes.

2. Toss a   Choose a   Outcomes
    coin   number

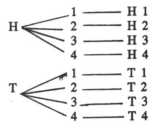

There are 8 possible outcomes.

3. Roll a   Choose a   Outcomes
    die   vowel

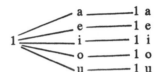

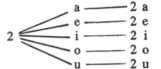

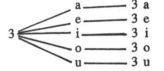

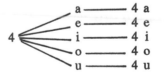

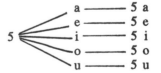

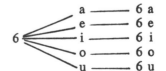

There are 30 possible outcomes.

4. Toss a   Choose a   Outcomes
    coin   number

$$
H
\begin{cases}
5 & \longrightarrow \; H\,5 \\
6 & \longrightarrow \; H\,6 \\
7 & \longrightarrow \; H\,7
\end{cases}
$$

$$
T
\begin{cases}
5 & \longrightarrow \; T\,5 \\
6 & \longrightarrow \; T\,6 \\
7 & \longrightarrow \; T\,7
\end{cases}
$$

There are 6 possible outcomes.

5. $\dfrac{\text{number of ways event can occur}}{\text{number of possible outcomes}}$

$= \dfrac{\text{number of white marbles}}{\text{total number of marbles}}$

$= \dfrac{4}{20}$

$= \dfrac{1}{5}$

6. $\dfrac{\text{number of pink marbles}}{\text{total number of marbles}} = \dfrac{3}{20}$

7. $\dfrac{\text{number of purple marbles}}{\text{total number of marbles}} = \dfrac{8}{20} = \dfrac{2}{5}$

8. $\dfrac{\text{number of orange marbles}}{\text{total number of marbles}} = \dfrac{5}{20} = \dfrac{1}{4}$

CHAPTER 8 SOLUTIONS TO PRACTICE TEST

1. Look for the largest sector. The 17–20 age group contains the most students.

2. amount = percent · base
            = 26% · 600
            = (0.26)(600)
            = 156

There were 156 students in the 26–30 age range.

3. The 31–40 age group contained 10% of the students.

4. (number of ○) · (value per ○) = 3 · $50
                                 = $150

5. Week 2

   (number of ○) · (value per ○) = 6 · $50
                                 = $300

6. Week 6

   1 ○ = $50

7. Week 1:   4 ($50) = $200
   Week 2:   6 ($50) = $300
   Week 3:   3 ($50) = $150
   Week 4:   3½ ($50) = $175
   Week 5:   2½ ($50) = $125
   Week 6:   1 ($50) = $50
      Total: $1000

A total of $1000 was taken in.

8. A (−4, −7)

9. B (0, 6)

10. C (−5, 5)

11. D (6, 0)

12.
$$
\begin{aligned}
x + 2y &= 1 \\
x + 2(0) &= 1 \\
x &= 1 \\
(1, &\, 0)
\end{aligned}
$$

$$
\begin{aligned}
x + 2y &= 1 \\
3 + 2y &= 1 \\
3 + 2y + (-3) &= 1 + (-3) \\
2y &= -2
\end{aligned}
$$

$$
\frac{2y}{2} = \frac{-2}{2}
$$

$$
\begin{aligned}
y &= -1 \\
(3, &\, -1)
\end{aligned}
$$

$$x + 2y = 1$$
$$-3 + 2y = 1$$
$$-3 + 2y + 3 = 1 + 3$$
$$2y = 4$$
$$\frac{2y}{2} = \frac{4}{2}$$
$$y = 2$$
$$(-3, 2)$$

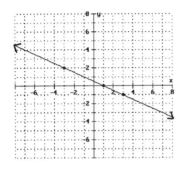

13.
$$y = 5x - 2$$
$$y = 5(0) - 2$$
$$y = -2$$
$$(0, -2)$$

$$y = 5x - 2$$
$$y = 5(1) - 2$$
$$y = 5 - 2$$
$$y = 3$$
$$(1, 3)$$

$$y = 5x - 2$$
$$-7 + 2 = 5x - 2 + 2$$
$$-5 = 5x$$
$$\frac{-5}{5} = \frac{5x}{5}$$
$$-1 = x$$
$$(-1, -7)$$

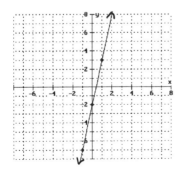

14. Let $x = 0$:
$$x - y = 2$$
$$0 - y = 2$$
$$-y = 2$$
$$\frac{-y}{-1} = \frac{2}{-1}$$
$$y = -2$$
$$(0, -2)$$

Let $y = 0$:
$$x - y = 2$$
$$x - 0 = 2$$
$$x = 2$$
$$(2, 0)$$

Let $y = 1$:
$$x - y = 2$$
$$x - 1 = 2$$
$$x - 1 + 1 = 2 + 1$$
$$x = 3$$
$$(3, 1)$$

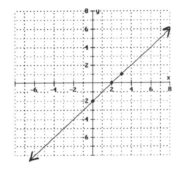

15.
$$x + 4 = 0$$
$$x + 4 + (-4) = 0 + (-4)$$
$$x = -4$$

This is a vertical line that crosses the x-axis at -4.

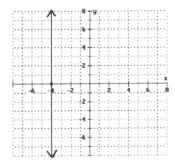

16. $y - 1 = 0$
    $y - 1 + 1 = 0 + 1$
        $y = 1$

This is a horizontal line that crosses the y-axis at 1.

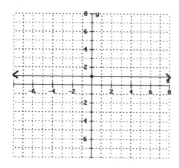

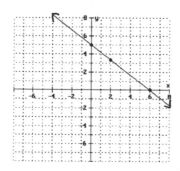

17. Let $x = 0$: $5x + 6y = 30$
                $5(0) + 6y = 30$
                     $6y = 30$

                $\dfrac{6y}{6} = \dfrac{30}{6}$

                     $y = 5$
                     $(0, 5)$

    Let $y = 0$: $5x + 6y = 30$
                $5x + 6(0) = 30$
                     $5x = 30$

                $\dfrac{5x}{5} = \dfrac{30}{5}$

                     $x = 6$
                     $(6, 0)$

    Let $x = 2$: $5x + 6y = 30$
                $5(2) + 6y = 30$
                $10 + 6y = 30$
        $10 + 6y + (-10) = 30 + (-10)$
                     $6y = 20$

                $\dfrac{6y}{6} = \dfrac{20}{6}$

                     $y = \dfrac{10}{3} = 3\dfrac{1}{3}$

                $\left(2, \dfrac{10}{3}\right)$

18. mean $= \dfrac{61 + 60 + 72 + 63 + 59 + 74}{6}$

          $= \dfrac{389}{6}$

          $= 64\dfrac{5}{6}$

    59, 60, $\underline{61}$, $\underline{63}$, 72, 74

    median $= \dfrac{61 + 63}{2}$

              $= \dfrac{124}{2}$

              $= 62$

    mode: There is none. (no number appears
           the most number of times)

19. mean $= \dfrac{101 + 94 + 113 + 89 + 126 + 94 + 112}{7}$

          $= \dfrac{729}{7}$

          $= 104.14$

    89, 94, 94, $\underline{101}$, 112, 113, 126

    median $= 101$

    mode $= 94$ (it appears the most number of times)

20.

Grade	Point Value of Grade	Credit Hours	(Point value) · (Credit Hours)
D	1	3	3
B	3	5	15
B	3	3	9
A	4	4	16
	Totals	15	43

$$\text{grade point average} = \frac{43}{15}$$

$$= 2.87$$

21. $\dfrac{\text{number of ways event can occur}}{\text{number of possible outcomes}} = \dfrac{2}{6} = \dfrac{1}{3}$

22. $\dfrac{\text{number of ways event can occur}}{\text{number of possible outcomes}} = \dfrac{1}{4}$

9.1 SOLUTIONS TO EXERCISES

1. angle

2. ray

3. line

4. line segment

5. measure of $\angle ABD$ = measure of $\angle ABC$ + measure of $\angle CBD$
   $$= 85° + 45°$$
   $$= 130°$$

6. measure of $\angle EBC$ = measure of $\angle EBD$ + measure of $\angle DBC$
   $$= 35° + 45°$$
   $$= 80°$$

7. measure of $\angle CBD = 45°$

8. acute angle

9. obtuse angle

10. right angle

11. the complement of a $64°$ angle $= 90° - 64° = 26°$

12. the supplement of an $88°$ angle $= 180° - 88° = 92°$

13. The $32°$ angle and $\angle y$ are vertical angles, hence measure of $\angle y = 32°$.

    $\angle x$ and $\angle y$ are adjacent angles, so the sum of their measures is $180°$. Measure of $\angle x = 180° - 32° = 148°$.

    $\angle x$ and $\angle z$ are vertical angles, hence measure of angle $\angle z = 148°$.

14. measure of $\angle x = 180° - 90° - 21° = 69°$

15. measure of $\angle x = 180° - 62° - 38° = 80°$

9.2 SOLUTIONS TO EXERCISES

1. 75 in. $= 75$ in. $\cdot \dfrac{1 \text{ ft}}{12 \text{ in.}}$

   $= \dfrac{75}{12}$ ft

   $= 6.25$ ft

2. 17 yd $= 17$ yd $\cdot \dfrac{3 \text{ ft}}{1 \text{ yd}}$

   $= \dfrac{51}{1}$ ft

   $= 51$ ft

3. $42{,}240$ ft $= 42{,}240$ ft $\cdot \dfrac{1 \text{ mile}}{5280 \text{ ft}}$

   $= \dfrac{42{,}240}{5280}$ miles

   $= 8$ miles

4. 2.6 mi $= 2.6$ mi $\cdot \dfrac{5280 \text{ ft}}{1 \text{ mi}}$
   $= 2.6 \cdot 5280$ ft
   $= 13{,}728$ ft

5. 82 ft $= 82$ ft $\cdot \dfrac{1 \text{ yd}}{3 \text{ ft}}$

   $= \dfrac{82}{3}$ yd

   $= 27\dfrac{1}{3}$ yd

6. 97 ft $= 97$ ft $\cdot \dfrac{12 \text{ in.}}{1 \text{ ft}}$

   $= 97 \cdot 12$ in.
   $= 1164$ in.

7.
$$
\begin{array}{r}
12 \text{ yd } 2 \text{ ft} \\
3\overline{)38} \\
\underline{3} \\
8 \\
\underline{6} \\
2
\end{array}
$$

8.
$$
\begin{array}{r}
6 \text{ ft } 11 \text{ in.} \\
12\overline{)83} \\
\underline{72} \\
11
\end{array}
$$

9.
$$
\begin{array}{r}
3 \text{ mi } 4160 \text{ ft} \\
5280\overline{)20000} \\
\underline{15840} \\
4160
\end{array}
$$

10. $6 \text{ ft} = 6 \text{ ft} \cdot \dfrac{12 \text{ in.}}{1 \text{ ft}}$

   $= 6 \cdot 12 \text{ in.}$

   $= 72 \text{ in.}$

   $6 \text{ ft. } 4 \text{ in.} = 72 \text{ in.} + 4 \text{ in.} = 76 \text{ in.}$

11. $12 \text{ yd} = 12 \text{ yd} \cdot \dfrac{3 \text{ ft}}{1 \text{ yd}}$

   $= 12 \cdot 3 \text{ ft}$

   $= 36 \text{ ft}$

   $12 \text{ yd } 2 \text{ ft} = 36 \text{ ft} + 2 \text{ ft} = 38 \text{ ft}$

12.   m   dm   cm
   _____↑
   2 units right

   65 m = 65.00
         _____↑
       = 6500 cm

13.   m   dm   cm   mm
   ↑_____
           3 units left

   5800 mm = 5800.
              ____↑
           = 5.8 m

14.   dm   cm   mm
   ↑_____
       2 units left

   19.6 mm = 19.6
              __↑

           = 0.196 dm

15.   m   dm   cm   mm
   _____↑
           3 units right

   0.9 m = 0.900
           ___↑
         = 900 mm

16.
$$
\begin{array}{r}
11 \text{ ft } 6 \text{ in.} \\
+ 8 \text{ ft } 9 \text{ in.} \\
\hline
19 \text{ ft } 15 \text{ in.}
\end{array}
$$
   $= 19 \text{ ft } 1 \text{ ft } 3 \text{ in.}$
   $= 20 \text{ ft } 3 \text{ in.}$

17.
$$
\begin{array}{r}
25 \text{ ft } 3 \text{ in.} = 24 \text{ ft } 15 \text{ in.} \\
- 10 \text{ ft } 8 \text{ in.} \quad - 10 \text{ ft } 8 \text{ in.} \\
\hline
14 \text{ ft } 7 \text{ in.}
\end{array}
$$

18.
$$
\begin{array}{r}
32 \text{ yd } 2 \text{ ft} \\
\times \qquad 5 \\
\hline
160 \text{ yd } 10 \text{ ft}
\end{array}
$$
   $= 160 \text{ yd } 3 \text{ yd } 1 \text{ ft}$
   $= 163 \text{ yd } 1 \text{ ft}$

19. 20 cm = 200 mm

$$
\begin{array}{r}
200 \text{ mm} \\
- 18 \text{ mm} \\
\hline
182 \text{ mm}
\end{array}
$$

   or 18 mm = 1.8 cm

$$
\begin{array}{r}
20.0 \text{ cm} \\
- 1.8 \text{ cm} \\
\hline
18.2 \text{ cm}
\end{array}
$$

20.
$$
\begin{array}{r}
3.2 \text{ m} \\
4\overline{)12.8} \\
\underline{12} \\
8 \\
\underline{8} \\
0
\end{array}
$$

9.3 SOLUTIONS TO EXERCISES

1. $P = 4s$
   $= 4(11 \text{ meters})$
   $= 44 \text{ meters}$

2. $P = a + b + c$
   $= 3 \text{ inches} + 10 \text{ inches} + 9 \text{ inches}$
   $= 22 \text{ inches}$

3. $P = 2l + 2w$
   $= 2(16 \text{ feet}) + 2(5 \text{ feet})$
   $= 32 \text{ feet} + 10 \text{ feet}$
   $= 42 \text{ feet}$

4. $P = 28 \text{ cm} + 28 \text{ cm} + 52 \text{ cm} + 52 \text{ cm}$
   $= 160 \text{ cm}$

5. $P = 4 \text{ ft} + 7 \text{ ft} + 2 \text{ ft} + 9 \text{ ft} + 13 \text{ ft}$
   $= 35 \text{ ft}$

6. $P = 29 \text{ m} + 5 \text{ m} + 20 \text{ m} + 6 \text{ m} + 9 \text{ m} + 11 \text{ m}$
   $= 80 \text{ meters}$

7. $P = 14 \text{ ft} + 18 \text{ ft} + 9 \text{ ft} + 7 \text{ ft} + 3 \text{ ft}$
   $= 51 \text{ ft}$

8. $P = a + b + c$
   $= 20 \text{ cm} + 30 \text{ cm} + 50 \text{ cm}$
   $= 100 \text{ cm}$

9. $P = 4s$
   $= 4(15 \text{ miles})$
   $= 60 \text{ miles}$

10. $P = 2l + 2w$
    $= 2(42 \text{ yd}) + 2(31 \text{ yd})$
    $= 84 \text{ yd} + 62 \text{ yd}$
    $= 146 \text{ yd}$

11. $P = 2l + 2w$
    $= 2(6 \text{ ft}) + 2(2 \text{ ft})$
    $= 12 \text{ ft} + 4 \text{ ft}$
    $= 16 \text{ ft}$

    16 feet of stripping is needed.

12. $P = 2l + 2w$
    $= 2(92 \text{ ft}) + 2(28 \text{ ft})$
    $= 184 \text{ ft} + 56 \text{ ft}$
    $= 240 \text{ ft}$

    240 feet of fencing is needed.

13. Let x = width of field
    then $4x$ = length of field

    $P = 2l + 2w$
    $600 = 2(4x) + 2(x)$
    $600 = 8x + 2x$
    $600 = 10x$

    $$\frac{600}{10} = \frac{10x}{10}$$

    $60 = x$

    The length of the field is
    $4x = 4(60 \text{ meters}) = 240 \text{ meters}$.

14. Let x = length of each equal side

    $P = a + b + c$
    $121 = x + x + 33$
    $121 = 2x + 33$
    $121 + (-33) = 2x + 33 + (-33)$
    $88 = 2x$

    $$\frac{88}{2} = \frac{2x}{2}$$

    $44 = x$

    The length of each equal side is 44 inches.

15. Let x = length
    then $2x - 12$ = width

    $P = 2l + 2w$
    $30 = 2(x) + 2(2x - 12)$
    $30 = 2x + 4x - 24$
    $30 = 6x - 24$
    $30 + 24 = 6x - 24 + 24$
    $54 = 6x$

    $$\frac{54}{6} = \frac{6x}{6}$$

    $9 = x$

    The width is $2x - 12$
    $= 2(9 \text{ meters}) - 12 \text{ meters} = 6 \text{ meters}$.

16. $P = 4s$
     $= 4(8 \text{ inches})$
     $= 32 \text{ inches}$

The perimeter is 32 inches.

9.4 SOLUTIONS TO EXERCISES

1. $A = lw$
     $= (4.8 \text{ feet})(2.1 \text{ feet})$
     $= 10.08 \text{ sq. feet}$

2. $A = \dfrac{1}{2}bh$

     $= \dfrac{1}{2}(5 \text{ cm})(3 \text{ cm})$

     $= \dfrac{15}{2} \text{ sq. cm}$

     or 7.5 sq. cm

3. $A = \dfrac{1}{2}bh$

     $= \dfrac{1}{2}(9 \text{ miles})(7 \text{ miles})$

     $= \dfrac{63}{2} \text{ sq. miles}$

     or 31.5 sq. miles

4. $A = bh$
     $= (10 \text{ meters})(4.5 \text{ meters})$
     $= 45 \text{ sq. meters}$

5. $A = \pi r^2$
              $= \pi (13 \text{ inches})^2$
              $= 169\pi \text{ sq. inches}$

    $169\pi \text{ sq. inches} \approx 169\left(\dfrac{22}{7}\right) \text{ sq. inches}$

                $= \dfrac{3718}{7} \text{ sq. inches}$

6. $d = 7 \text{ feet}$

         $r = \dfrac{1}{2}d$

            $= \dfrac{1}{2}(7 \text{ feet})$

            $= \dfrac{7}{2} \text{ feet}$

       $A = \pi r^2$

            $= \pi \left(\dfrac{7}{2} \text{ feet}\right)^2$

            $= \dfrac{49}{4}\pi \text{ sq. feet}$

   $\dfrac{49}{4}\pi \text{ sq. feet} \approx \dfrac{49}{4}(3.14) \text{ sq. feet}$

               $= 38.465 \text{ sq. feet}$

7. $A = \dfrac{1}{2}(b + B)h$

     $= \dfrac{1}{2}(7 \text{ in.} + 12 \text{ in.})(2 \text{ in.})$

     $= \dfrac{1}{2}(19 \text{ in.})(2 \text{ in.})$

     $= 19 \text{ sq. in.}$

8. $A = \dfrac{1}{2}(b + B)h$

     $= \dfrac{1}{2}(5 \text{ cm} + 11 \text{ cm})(7 \text{ cm})$

     $= \dfrac{1}{2}(16 \text{ cm})(7 \text{ cm})$

     $= 56 \text{ sq. cm}$

9.

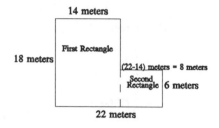

A = area of   +   area of

    1st rectangle   2nd rectangle

= (18 meters)(14 meters) + (8 meters)(6 meters)
= 252 sq. meters + 48 sq. meters
= 300 sq. meters

10. $V = lwh$
    = (2 in.)(8 in.)(3 in.)
    = 48 cu. in.

11. $V = \dfrac{4}{3}\pi r^3$

    $= \dfrac{4}{3}\pi (28 \text{ miles})^3$

    $= \dfrac{87808}{3}\pi$ cu. miles

    $\approx \dfrac{87808}{3}(3.14)$ cu. miles

    = 91905.707 cu. miles

12. $V = \pi r^2 h$
    $= \pi (2.2 \text{ feet})^2 (16 \text{ feet})$
    $= \pi (4.84 \text{ sq. feet})(16 \text{ feet})$
    $= 77.44\pi$ cu. feet
    $\approx 77.44(3.14)$ cu. feet
    = 243.1616 cu. feet

13. $V = \dfrac{1}{3}\pi r^2 h$

    $= \dfrac{1}{3}\pi (5 \text{ meters})^2 (19 \text{ meters})$

    $= \dfrac{1}{3}\pi (25 \text{ sq. meters})(19 \text{ meters})$

$= \dfrac{475}{3}\pi$ cu. meters

$\approx \dfrac{475}{3}(3.14)$ cu. meters

= 497.167 cu. meters

14. $V = \dfrac{1}{3}s^2 h$

    $= \dfrac{1}{3}(4.7 \text{ in.})^2 (18.6 \text{ in.})$

    $= \dfrac{1}{3}(22.09 \text{ sq. in})(18.6 \text{ in.})$

    = 136.958 cu. in.

15. $A = lw$

    $= (20 \text{ feet})\left(14\dfrac{1}{2} \text{ feet}\right)$

    $= (20 \text{ feet})\left(\dfrac{29}{2} \text{ feet}\right)$

    = 290 sq. feet
The area of the wall is 290 sq. feet.

16. $A = lw$

    $= (14 \text{ in.})\left(8\dfrac{1}{2} \text{ in.}\right)$

    $= (14 \text{ in.})\left(\dfrac{17}{2} \text{ in.}\right)$

    = 119 sq. in.
The frame requires 119 sq. in. of glass.

17. $V = s^3$

    $= \left(2\dfrac{3}{4} \text{ in.}\right)^3$

    $= \left(\dfrac{11}{4} \text{ in.}\right)^3$

    $= \dfrac{1331}{64}$ cu. in.
    ≈ 20.797 cu. in.

18. $V = lwh$

$= (3 \text{ ft})\left(1\frac{1}{2}\text{ ft}\right)\left(\frac{1}{2}\text{ ft}\right)$

$= (3 \text{ ft})\left(\frac{3}{2}\text{ ft}\right)\left(\frac{1}{2}\text{ ft}\right)$

$= \frac{9}{4}$ cu. ft

$= 2.25$ cu. ft

The volume of the block of ice is 2.25 cu. ft.

9.5 SOLUTIONS TO EXERCISES

1. $5 \text{ lb} = 5 \text{ lb} \cdot \frac{16 \text{ oz}}{1 \text{ lb}}$

$= 5 \cdot 16 \text{ oz}$
$= 80 \text{ oz}$

2. $6 \text{ tons} = 6 \text{ tons} \cdot \frac{2000 \text{ lb}}{1 \text{ ton}}$

$= 6 \cdot 2000 \text{ lb}$
$= 12,000 \text{ lb}$

3. $7200 \text{ lb} = 7200 \text{ lb} \cdot \frac{1 \text{ ton}}{2000 \text{ lb}}$

$= \frac{7200}{2000}$ tons

$= 3.6 \text{ tons}$

4. $22.25 \text{ lb} = 22.25 \text{ lb} \cdot \frac{16 \text{ oz}}{1 \text{ lb}}$

$= 22.25 \cdot 16 \text{ oz}$
$= 356 \text{ oz}$

5. $82 \text{ oz} = 82 \text{ oz} \cdot \frac{1 \text{ lb}}{16 \text{ oz}}$

$= \frac{82}{16}$ lb

$= 5.1 \text{ lb}$

6. kg   hg   dag   g

    ↑ ‾‾‾‾‾‾‾‾‾‾‾‾‾‾
       3 units left

    6000 g = 6000.
             ↑‾‾‾
    = 6.0 kg

7. g   dg   cg   mg

    ‾‾‾‾‾‾‾‾‾‾‾‾‾↑
       3 units right

    12 g = 12.000
              ‾‾↑
    = 12000 mg

8. kg   hg   dag   g

    ↑ ‾‾‾‾‾‾‾‾‾‾‾‾‾
       3 units left

    7.2 g = 007.2
            ↑‾
    = 0.0072 kg

9. kg   hg   dag   g

    ‾‾‾‾‾‾‾‾‾‾‾‾‾↑
       3 units right

    6.92 kg = 6.920
                ‾↑
    = 6920 g

10. hg   dag   g   dg   cg

    ↑ ‾‾‾‾‾‾‾‾‾‾‾‾‾‾‾‾‾
       4 units left

    8025 cg = 8025.
              ↑‾‾
    = 0.8025 hg

11.   19 lb   4 oz
    + 31 lb 15 oz
      ‾‾‾‾‾‾‾‾‾‾‾‾‾
      50 lb 19 oz
    = 50 lb   1 lb 3 oz
    = 51 lb   3 oz

12.   1 ton   1905 lb
    + 5 tons   168 lb
      ‾‾‾‾‾‾‾‾‾‾‾‾‾‾‾‾
      6 tons 2073 lb
    = 6 tons 1 ton 73 lb
    = 7 tons   73 lb

13. $\begin{array}{r} 14 \text{ lb } 3 \text{ oz} = 13 \text{ lb } 19 \text{ oz} \\ - 4 \text{ lb } 8 \text{ oz} - 4 \text{ lb } 8 \text{ oz} \\ \hline 9 \text{ lb } 11 \text{ oz} \end{array}$

14. $\begin{array}{r} 3 \text{ lb } 7 \text{ oz} \\ \times \quad 9 \\ \hline 27 \text{ lb } 63 \text{ oz} \end{array}$
   = 27 lb   3 lb 15 oz
   = 30 lb 15 oz

15. $\begin{array}{r} 1 \text{ ton} \quad 460 \text{ lb} \\ 5\overline{)6 \text{ tons}} \quad 300 \text{ lb} \\ \underline{5} \end{array}$
    1 ton = 2000 lb
              2300 lb

16. $\begin{array}{r} 89.7 \text{ g} \\ + 10.6 \text{ g} \\ \hline 100.3 \text{ g} \end{array}$

17. 6 kg = 6000 g

      $\begin{array}{r} 6000 \text{ g} \\ - 3560 \text{ g} \\ \hline 2440 \text{ g} \end{array}$
   or   3560 g = 3.56 kg

      $\begin{array}{r} 6.00 \text{ kg} \\ - 3.56 \text{ kg} \\ \hline 2.44 \text{ kg} \end{array}$

18. $\begin{array}{r} 2.9 \text{ kg} \\ \times 8.6 \\ \hline 174 \\ 232 \\ \hline 24.94 \text{ kg} \end{array}$

19. Let x = weight of 9 cans

   $\dfrac{1 \text{ can}}{15 \text{ oz}} = \dfrac{9 \text{ cans}}{x \text{ oz}}$

   $1 \cdot x = 15 \cdot 9$
   $ x = 135$

      $\begin{array}{r} 8 \text{ lb } 7 \text{ oz} \\ 16\overline{)135} \\ \underline{128} \\ 7 \end{array}$

   9 cans weigh 135 oz or 8 lb 7 oz.

20. 20000 g = 20 kg

      $\begin{array}{r} 105 \text{ kg} \\ - 20 \text{ kg} \\ \hline 85 \text{ kg} \end{array}$

   His new weight is 85 kg.

9.6 SOLUTIONS TO EXERCISES

1. 28 qt = 28 qt $\cdot \dfrac{1 \text{ gal}}{4 \text{ qt}}$

   $= \dfrac{28}{4}$ gal

   = 7 gal

2. 11 qt = 11 qt $\cdot \dfrac{2 \text{ pt}}{1 \text{ qt}}$

   $= 11 \cdot 2$ pt

   = 22 pt

3. 18 c = 18 c $\cdot \dfrac{1 \text{ pt}}{2 \text{ c}}$

   $= \dfrac{18}{2}$ pt

   = 9 pt

4. 40 c = 40 c $\cdot \dfrac{1 \text{ pt}}{2 \text{ c}} \cdot \dfrac{1 \text{ qt}}{2 \text{ pt}} \cdot \dfrac{1 \text{ gal}}{4 \text{ qt}}$

   $= \dfrac{40}{16}$ gal

   $= 2\dfrac{1}{2}$ gal

5. $5\dfrac{3}{4}$ qt $= \dfrac{23}{4}$ qt $\cdot \dfrac{2 \text{ pt}}{1 \text{ qt}} \cdot \dfrac{2 \text{ c}}{1 \text{ pt}}$

   = 23 c

6. $5\dfrac{7}{8}$ gal $= \dfrac{47}{8}$ gal $\cdot \dfrac{4 \text{ qt}}{1 \text{ gal}} \cdot \dfrac{2 \text{ pt}}{1 \text{ qt}}$

   = 47 pt

7. L   dl   cl   ml
    _____↑
         3 units right

    9 L = 9.000
            ___↑
        = 9000 ml

8. L   dl   cl   ml
    ↑_____
         3 units left

    3200 ml = 3200.
                 ___↑
            = 3.2 L

9. kl   hl   dal   L
    ↑_____
          3 units left

    140 L = 140.
              ___↑
          = 0.14 kl

10. kl   hl   dal   L
    _____↑
          3 units right

    2.5 = 2.500
            ___↑
        = 2500 L

11.   8 gal   2 qt
     + 6 gal   1 qt
      14 gal   3 qt

12.   7 c   4 fl oz
     + 3 c   6 fl oz
      10 c   10 fl oz

     = 10 c   1 c 2 fl oz
     = 11 c   2 fl oz

13.   4 gal   1 pt
     ×   3
      12 gal   3 pt
     = 12 gal   1 qt   1 pt

14.   3 pt   = 2 pt   2 c
     − 2 pt   1 c   − 2 pt   1 c
                              1 c

15.   20.2 L
     + 14.6 L
       34.8 L

16. 7920 ml = 7.920 L

       7.92 L
     − 0.2 L
       7.72 L

     or   0.2 L = 200 ml

       7920 ml
     − 200 ml
       7720 ml

                    8.
17. 0.8.)6.4.
            64
             0

     6.4 L ÷ 0.8 = 8 L

18.   125 ml
     ×   7
       875 ml

19. $1\frac{1}{2}$ c = $\frac{3}{2}$c · $\frac{8 \text{ fl oz}}{1 \text{ c}}$

         = 12 fl oz

20.   1 L = 1000 ml

     $\frac{1000 \text{ ml}}{4 \text{ children}}$ = 250 ml/child

     Each child will get 250 ml.

9.7 SOLUTIONS TO EXERCISES

1. F = $\frac{9C}{5}$ + 32

      = $\frac{9(85)}{5}$ + 32

      = 153 + 32
      = 185

    185°F

2. $F = \dfrac{9C}{5} + 32$

$= \dfrac{9(30)}{5} + 32$

$= 54 + 32$
$= 86$

$86°F$

3. $C = \dfrac{5(F - 32)}{9}$

$= \dfrac{5(113 - 32)}{9}$

$= \dfrac{5(81)}{9}$

$= 45$

$45°C$

4. $C = \dfrac{5(F - 32)}{9}$

$= \dfrac{5(77 - 32)}{9}$

$= \dfrac{5(45)}{9}$

$= 25$

$25°C$

5. $F = \dfrac{9C}{5} + 32$

$= \dfrac{9(105)}{5} + 32$

$= 189 + 32$
$= 221$

$221°F$

6. $C = \dfrac{5(F - 32)}{9}$

$= \dfrac{5(58 - 32)}{9}$

$= \dfrac{5(26)}{9}$

$= 14.4$

$14.4°C$

7. $C = \dfrac{5(F - 32)}{9}$

$= \dfrac{5(49 - 32)}{9}$

$= \dfrac{5(17)}{9}$

$= 9.4$

$9.4°C$

8. $F = \dfrac{9C}{5} + 32$

$= \dfrac{9(80)}{5} + 32$

$= 144 + 32$
$= 176$

$176°F$

9. $F = \dfrac{9C}{5} + 32$

$= \dfrac{9(22.4)}{5} + 32$

$= 40.32 + 32$
$= 72.32$

$72.3°F$

10. $C = \dfrac{5(F - 32)}{9}$

$= \dfrac{5(121.7 - 32)}{9}$

$= \dfrac{5(89.7)}{9}$

$= 49.8$

$49.8°C$

11. $C = \dfrac{5(F - 32)}{9}$

$= \dfrac{5(95 - 32)}{9}$

$= \dfrac{5(63)}{9}$

$= 35$

$35°C$

12. $C = \dfrac{5(F - 32)}{9}$

$= \dfrac{5(70 - 32)}{9}$

$= \dfrac{5(38)}{9}$

$= 21.1$

$21.1°C$

13. $F = \dfrac{9C}{5} + 32$

$= \dfrac{9(20)}{5} + 32$

$= 36 + 32$
$= 68$

$68°F$

14. $F = \dfrac{9C}{5} + 32$

$\quad = \dfrac{9(12)}{5} + 32$

$\quad = 21.6 + 32$

$\quad = 53.6$

$\quad 53.6°F$

15. $C = \dfrac{5(F - 32)}{9}$

$\quad = \dfrac{5(102.3 - 32)}{9}$

$\quad = \dfrac{5(70.3)}{9}$

$\quad = 39.1$

$\quad 39.1°C$

CHAPTER 9 PRACTICE TEST SOLUTIONS

1. complement: $90° - 57° = 33°$
   supplement: $180° - 57° = 123°$

2. The 70° angle and $\angle y$ are vertical angles, hence the measure of $\angle y = 70°$.

   $\angle x$ and $\angle y$ are adjacent angles. The measure of $\angle x = 180° - 70° = 110°$.

   $\angle x$ and $\angle z$ are vertical angles. The measure of $\angle z$ is 110°.

3. measure of $\angle x = 180° - 90° - 50° = 40°$

4. $P = 2l + 2w$
   $\quad = 2(30) + 2(19)$
   $\quad = 60 + 38$
   $\quad = 98$

   The perimeter is 98 feet.

5. $\quad$ Let x = width,
   then $x + 12$ = length.

   $P = 2l + 2w$
   $120 = 2(x + 12) + 2(x)$
   $120 = 2x + 24 + 2x$
   $120 = 4x + 24$
   $96 = 4x$

   $\dfrac{96}{4} = \dfrac{4x}{4}$

   $24 = x$
   $x + 12 = 24 + 12 = 36$

   The width is 24 inches and the length is 36 inches.

6. $A = lw$
   $\quad = (11 \text{ inches})(4.5 \text{ inches})$
   $\quad = 49.5 \text{ sq. inches}$

7. $V = s^3$

   $\quad = \left(8\dfrac{1}{4} \text{ meters}\right)^3$

   $\quad = \left(\dfrac{33}{4} \text{ meters}\right)^3$

   $\quad = 561.516 \text{ cu. meters}$

8. $V = lwh$

   $\quad = (3 \text{ ft})\left(1\dfrac{1}{2} \text{ ft}\right)\left(\dfrac{1}{2} \text{ ft}\right)$

   $\quad = (3 \text{ ft})\left(\dfrac{3}{2} \text{ ft}\right)\left(\dfrac{1}{2} \text{ ft}\right)$

   $\quad = \dfrac{9}{4} \text{ cu. ft}$

   $\quad = 2.25 \text{ cu. ft}$

   It will take 2.25 cu. ft of potting soil.

9. $\quad \begin{array}{r} 11 \text{ ft } 8 \text{ in.} \\ 12\overline{)140} \\ \underline{12} \\ 20 \\ \underline{12} \\ 8 \end{array}$

10. $26 \text{ qt} = 26 \text{ qt} \cdot \dfrac{1 \text{ gal}}{4 \text{ qt}}$

       $= \dfrac{26}{4} \text{ gal}$

       $= 6\dfrac{1}{2} \text{ gal}$

11. $46 \text{ oz} = 46 \text{ oz} \cdot \dfrac{1 \text{ lb}}{16 \text{ oz}}$

       $= \dfrac{46}{16} \text{ lb}$

       $= 2\dfrac{7}{8} \text{ lb}$

12. $4.6 \text{ ton} = 4.6 \text{ ton} \cdot \dfrac{2000 \text{ lb}}{1 \text{ ton}}$

       $= 4.6 \cdot 2000 \text{ lb}$
       $= 9200 \text{ lb}$

13. $42 \text{ pt} = 42 \text{ pt} \cdot \dfrac{1 \text{ qt}}{2 \text{ pt}} \cdot \dfrac{1 \text{ gal}}{4 \text{ qt}}$

       $= \dfrac{42}{8} \text{ gal}$

       $= 5\dfrac{1}{4} \text{ gal}$

14.   g   dg   cg   mg
     ↑
     3 units left

     62 mg = 062.
           ↑

       = 0.062 g

15.   kg   hg   dag   g
                 ↑
     3 units right

     9.8 kg = 9.800
             ↑

       = 9800 g

16.   cm   mm
               ↑
     1 unit right

     7.2 cm = 7.2
             ↑

       = 72 mm

17.   g   dg
     ↑
     1 unit left

     9.1 dg = 9.1
         ↑

       = 0.91 g

18.   L   dl   cl   ml
                ↑
     3 units right

     0.075 L = 0.075
               ↑

       = 75 ml

19.   5 qt   1 pt
      + 7 qt   1 pt
       12 qt   2 pt

     = 12 qt   1 qt
     = 13 qt

20.   10 lb   5 oz   =   9 lb   21 oz
     − 3 lb   7 oz   =   −3 lb   7 oz
                      6 lb   14 oz

21.   3 ft   4 in.
      ×   5
      15 ft   20 in.

     = 15 ft   1 ft   8 in.
     = 16 ft   8 in.

22.
$$
\begin{array}{r}
2 \text{ gal} \quad 3 \text{ qt} \\
3\overline{)8 \text{ gal}} \quad 1 \text{ qt} \\
\underline{6} \qquad\qquad
\end{array}
$$

       $2 \text{ gal} = \dfrac{8 \text{ qt}}{9 \text{ qt}}$

23.   11 cm = 110 mm

     110 mm
     − 16 mm
      94 mm

    or 16 mm = 1.6 cm

     11.0 cm
     − 1.6 cm
      9.4 cm

24.   2.4 km = 2400 m

     2400 m
     + 329 m
     2729 m

    or 329 m = 0.329 km

     2.4   km
     + 0.329 km
     2.729 km

25. $C = \dfrac{5(F - 32)}{9}$

     $= \dfrac{5(74 - 32)}{9}$

     $= \dfrac{5(42)}{9}$

     $= 23.3$

    23.3°C

26. $F = \dfrac{9C}{5} + 32$

     $= \dfrac{9(10.8)}{5} + 32$

     $= 19.44 + 32$
     $= 51.44$

    51.44°F

10.1 SOLUTIONS TO EXERCISES

1. $(3x - 2) + (-9x + 31) = (3x - 9x) + (-2 + 31)$
$$= -6x + 29$$

2. $(18y + 6) + (10y - 22)$
$= (18y + 10y) + (6 - 22)$
$= 28y - 16$

3. $(2t + 9) + (7t^2 - 8t + 5)$
$= 7t^2 + (2t - 8t) + (9 + 5)$
$= 7t^2 - 6t + 14$

4. $(30x + 5) - (25x - 2)$
$= (30x + 5) + (-25x + 2)$
$= (30x - 25x) + (5 + 2)$
$= 5x + 7$

5. $(-18x^2 + 4x - 1) - (-2x + 17)$
$= (-18x^2 + 4x - 1) + (2x - 17)$
$= -18x^2 + (4x + 2x) + (-1 - 17)$
$= -18x^2 + 6x - 18$

6. $(10y^3 + 15y^2 - y - 2) - (8y^3 + 6y - 12)$
$= (10y^3 + 15y^2 - y - 2) + (-8y^3 - 6y + 12)$
$= (10y^3 - 8y^3) + 15y^2 + (-y - 6y) + (-2 + 12)$
$= 2y^3 + 15y^2 - 7y + 10$

7. $(3.7b^3 + 20) + (-5.3b^2 - 6.1b + 7)$
$= 3.7b^3 - 5.3b^2 - 6.1b + (20 + 7)$
$= 3.7b^3 - 5.3b^2 - 6.1b + 27$

8. $\begin{array}{r} -9t - 7 \\ -\ (16t - 7) \\ \hline \end{array}$

$\begin{array}{r} -9t - 7 \\ -\ 16t + 7 \\ \hline -25t \end{array}$

9. $\begin{array}{r} 21x^2 + 5x - 5 \\ -\ (\ x^2 - 6x + 5) \\ \hline \end{array}$

$\begin{array}{r} 21x^2 + 5x - 5 \\ -\ x^2 + 6x - 5 \\ \hline 20x^2 + 11x - 10 \end{array}$

10. $\begin{array}{r} -3x^2 + \dfrac{5}{8}x \\ -\left(4x^2 - \dfrac{3}{8}x\right) \\ \hline \end{array}$

$\begin{array}{r} -3x^2 + \dfrac{5}{8}x \\ -\ 4x^2 + \dfrac{3}{8}x \\ \hline -7x^2 + \ x \end{array}$

11. $-9x - 2 = -9(3) - 2$
$$= -27 - 2$$
$$= -29$$

12. $4x + 13 = 4(3) + 13$
$$= 12 + 13$$
$$= 25$$

13. $x^2 - 2x + 5 = (3)^2 - 2(3) + 5$
$$= 9 - 6 + 5$$
$$= 8$$

14. $-x^2 - x - 1 = -(3)^2 - 3 - 1$
$$= -9 - 3 - 1$$
$$= -13$$

15. $\dfrac{4x^2}{6} + 10 = \dfrac{4(3)^2}{6} + 10$

$$= \dfrac{4(9)}{6} + 10$$

$$= \dfrac{36}{6} + 10$$

$$= 6 + 10$$
$$= 16$$

16. $\dfrac{x^3}{9} - x - 11 = \dfrac{3^3}{9} - 3 - 11$

$$= \dfrac{27}{9} - 3 - 11$$

$$= 3 - 3 - 11$$
$$= -11$$

17. $4x - 1 = 4(-4) - 1$
$ = -16 - 1$
$ = -17$

18. $x^2 = (-4)^2$
$ = 16$

19. $x^3 = (-4)^3$
$ = -64$

20. $x^3 - x^2 + x + 1 = (-4)^3 - (-4)^2 + (-4) + 1$
$ = -64 - 16 + (-4) + 1$
$ = -83$

10.2 SOLUTIONS TO EXERCISES

1. $x^4 \cdot x^9 = x^{4+9}$
$ = x^{13}$

2. $b^{12} \cdot b = b^{12+1}$
$\phantom{b^{12} \cdot b} = b^{13}$

3. $2y^3 \cdot 6y^4 = (2 \cdot 6)(y^3 \cdot y^4)$
$ = 12y^{3+4}$
$ = 12y^7$

4. $-8x \cdot 11x = (-8 \cdot 11)(x \cdot x)$
$ = -88x^{1+1}$
$ = -88x^2$

5. $(-10a^2b)(-2a^6b^5) = (-10)(-2)(a^2 \cdot a^6)(b \cdot b^5)$
$ = 20a^{2+6}b^{1+5}$
$ = 20a^8b^6$

6. $(-x^3y^5z)(-3xy^2z^4) = (-1)(-3)(x^3 \cdot x)(y^5 \cdot y^2)(z \cdot z^4)$
$ = 3x^{3+1}y^{5+2}z^{1+4}$
$ = 3x^4y^7z^5$

7. $2x \cdot 6x \cdot x^2 = (2 \cdot 6)(x \cdot x \cdot x^2)$
$ = 12x^{1+1+2}$
$ = 12x^4$

8. $a \cdot 9a^{10} \cdot 10a^9 = (9 \cdot 10)(a \cdot a^{10} \cdot a^9)$
$\phantom{a \cdot 9a^{10} \cdot 10a^9} = 90a^{20}$

9. $(x^{10})^9 = x^{10 \cdot 9}$
$\phantom{(x^{10})^9} = x^{90}$

10. $(y^{15})^3 = y^{15 \cdot 3}$
$\phantom{(y^{15})^3} = y^{45}$

11. $(2m)^4 = 2^4 m^4$
$ = 16m^4$

12. $(y^6)^3 \cdot (y^2)^4 = y^{6 \cdot 3} \cdot y^{2 \cdot 4}$
$ = y^{18} \cdot y^8$
$ = y^{18+8}$
$ = y^{26}$

13. $(x^3y^8)^7 = (x^3)^7(y^8)^7$
$ = x^{3 \cdot 7}y^{8 \cdot 7}$
$ = x^{21}y^{56}$

14. $(8m^5n^{12})^3 = 8^3(m^5)^3(n^{12})^3$
$\phantom{(8m^5n^{12})^3} = 512m^{5 \cdot 3}n^{12 \cdot 3}$
$\phantom{(8m^5n^{12})^3} = 512m^{15}n^{36}$

15. $(-4z)(2z^9)^4 = -4 \cdot z \cdot 2^4 \cdot (z^9)^4$
$ = -4 \cdot z \cdot 16 \cdot z^{9 \cdot 4}$
$ = -4 \cdot z \cdot 16 \cdot z^{36}$
$ = -4 \cdot 16 \cdot z^{1+36}$
$ = -64z^{37}$

16. $(2xy^2)^5(3x^6y^4)^3$
$= 2^5x^5(y^2)^5 \cdot 3^3(x^6)^3(y^4)^3$
$= 32x^5y^{2 \cdot 5} \cdot 27x^{6 \cdot 3}y^{4 \cdot 3}$
$= 32x^5y^{10} \cdot 27x^{18}y^{12}$
$= (32 \cdot 27)(x^5 \cdot x^{18})(y^{10} \cdot y^{12})$
$= 864x^{5+18}y^{10+12}$
$= 864x^{23}y^{22}$

17. $(14y^{12}z^8)^2 = 14^2(y^{12})^2(z^8)^2$
$\phantom{(14y^{12}z^8)^2} = 196y^{12 \cdot 2}z^{8 \cdot 2}$
$\phantom{(14y^{12}z^8)^2} = 196y^{24}z^{16}$

18. $(4a^3b^{12})^3(2a^7b)^4$

$= 4^3(a^3)^3(b^{12})^3 \cdot 2^4(a^7)^4b^4$

$= 64a^{3\cdot3}b^{12\cdot3} \cdot 16a^{7\cdot4}b^4$

$= 64a^9b^{36} \cdot 16a^{28}b^4$

$= (64 \cdot 16)(a^9 \cdot a^{28})(b^{36} \cdot b^4)$

$= 1024a^{9+28}b^{36+4}$

$= 1024a^{37}b^{40}$

19. $A = s^2$

$= (3x^5 \text{ feet})^2$

$= 3^2(x^5)^2 \text{ sq. feet}$

$= 9x^{5\cdot2} \text{ sq.feet}$

$= 9x^{10} \text{ sq. feet}$

20. $A = l \cdot w$

$= (2y^4 \text{ meters}) \cdot (2y \text{ meters})$

$= 2 \cdot 2 \cdot y^4 \cdot y \text{ sq. meters}$

$= 4y^{4+1} \text{ sq. meters}$

$= 4y^5 \text{ sq. meters}$

10.3 SOLUTIONS TO EXERCISES

1. $7x(8x^2 + 5) = 7x \cdot 8x^2 + 7x \cdot 5$
$= 56x^3 + 35x$

2. $3y(9y^4 - y) = 3y \cdot 9y^4 + 3y(-y)$
$= 27y^5 - 3y^2$

3. $-6m(3m^2 - m - 2)$
$= -6m \cdot 3m^2 - 6m(-m) - 6m(-2)$
$= -18m^3 + 6m^2 + 12m$

4. $8t^3(-2t^2 + t + 6)$
$= 8t^3(-2t^2) + 8t^3 \cdot t + 8t^3 \cdot 6$
$= -16t^5 + 8t^4 + 48t^3$

5. $(x - 3)(x - 9) = x(x - 9) - 3(x - 9)$
$= x \cdot x + x(-9) - 3 \cdot x - 3(-9)$
$= x^2 - 9x - 3x + 27$
$= x^2 - 12x + 27$

6. $(y + 10)(y + 1)$
$= y(y + 1) + 10(y + 1)$
$= y \cdot y + y \cdot 1 + 10 \cdot y + 10 \cdot 1$
$= y^2 + y + 10y + 10$
$= y^2 + 11y + 10$

7. $(m + 8)(m - 8)$
$= m(m - 8) + 8(m - 8)$
$= m \cdot m + m(-8) + 8 \cdot m + 8(-8)$
$= m^2 - 8m + 8m - 64$
$= m^2 - 64$

8. $(3x + 5)(x - 11)$
$= 3x(x - 11) + 5(x - 11)$
$= 3x \cdot x + 3x(-11) + 5 \cdot x + 5(-11)$
$= 3x^2 - 33x + 5x - 55$
$= 3x^2 - 28x - 55$

9. $(2t + 5)(2t + 7)$
$= 2t(2t + 7) + 5(2t + 7)$
$= 2t \cdot 2t + 2t \cdot 7 + 5 \cdot 2t + 5 \cdot 7$
$= 4t^2 + 14t + 10t + 35$
$= 4t^2 + 24t + 35$

10. $(4y - 5)^2 = (4y - 5)(4y - 5)$
$= 4y(4y - 5) - 5(4y - 5)$
$= 4y \cdot 4y + 4y(-5) - 5(4y) - 5(-5)$
$= 16y^2 - 20y - 20y + 25$
$= 16y^2 - 40y + 25$

11. $(3x + 8)^2 = (3x + 8)(3x + 8)$
$= 3x(3x + 8) + 8(3x + 8)$
$= 3x \cdot 3x + 3x \cdot 8 + 8 \cdot 3x + 8 \cdot 8$
$= 9x^2 + 24x + 24x + 64$
$= 9x^2 + 48x + 64$

12. $\left(z + \dfrac{1}{4}\right)\left(z - \dfrac{3}{4}\right)$

$= z\left(z - \dfrac{3}{4}\right) + \dfrac{1}{4}\left(z - \dfrac{3}{4}\right)$

$= z \cdot z + z\left(-\dfrac{3}{4}\right) + \dfrac{1}{4} \cdot z + \dfrac{1}{4}\left(-\dfrac{3}{4}\right)$

$= z^2 - \dfrac{3}{4}z + \dfrac{1}{4}z - \dfrac{3}{16}$

$= z^2 - \dfrac{1}{2}z - \dfrac{3}{16}$

13. $(4a + 9)(4a - 9)$
$= 4a(4a - 9) + 9(4a - 9)$
$= 4a \cdot 4a + 4a(-9) + 9 \cdot 4a + 9(-9)$
$= 16a^2 - 36a + 36a - 81$
$= 16a^2 - 81$

14. $(x + 7)^2 = (x + 7)(x + 7)$
$= x(x + 7) + 7(x + 7)$
$= x \cdot x + x \cdot 7 + 7 \cdot x + 7 \cdot 7$
$= x^2 + 7x + 7x + 49$
$= x^2 + 14x + 49$

15. $(a - 2)(6a^2 + 5a + 13)$
$= a(6a^2 + 5a + 13) - 2(6a^2 + 5a + 13)$
$= a \cdot 6a^2 + a \cdot 5a + a \cdot 13 - 2 \cdot 6a^2 - 2 \cdot 5a$
$\quad - 2 \cdot 13$
$= 6a^3 + 5a^2 + 13a - 12a^2 - 10a - 26$
$= 6a^3 - 7a^2 + 3a - 26$

16   $(x + 5)(x^2 - 5x + 25)$
$= x(x^2 - 5x + 25) + 5(x^2 - 5x + 25)$
$= x \cdot x^2 + x(-5x) + x \cdot 25 + 5 \cdot x^2 + 5(-5x)$
$\quad + 5 \cdot 25$
$= x^3 - 5x^2 + 25x + 5x^2 - 25x + 125$
$= x^3 + 125$

17. $(y^2 + 3)(2y^2 - y + 8)$
$= y^2(2y^2 - y + 8) + 3(2y^2 - y + 8)$
$= y^2 \cdot 2y^2 + y^2(-y) + y^2 \cdot 8 + 3 \cdot 2y^2 + 3(-y)$
$\quad + 3 \cdot 8$
$= 2y^4 - y^3 + 8y^2 + 6y^2 - 3y + 24$
$= 2y^4 - y^3 + 14y^2 - 3y + 24$

18. $(x^2 + 2x + 3)(x^3 - x^2 + 4)$
$= x^2(x^3 - x^2 + 4) + 2x(x^3 - x^2 + 4)$
$\quad + 3(x^3 - x^2 + 4)$
$= x^2 \cdot x^3 + x^2(-x^2) + x^2 \cdot 4 + 2x \cdot x^3 + 2x(-x^2)$
$\quad + 2x \cdot 4 + 3 \cdot x^3 + 3(-x^2) + 3 \cdot 4$
$= x^5 - x^4 + 4x^2 + 2x^4 - 2x^3 + 8x + 3x^3$
$\quad - 3x^2 + 12$
$= x^5 + x^4 + x^3 + x^2 + 8x + 12$

19. $A = s^2$
$= (3x + 10)^2$
$= (3x + 10)(3x + 10)$
$= 3x(3x + 10) + 10(3x + 10)$
$= 3x \cdot 3x + 3x \cdot 10 + 10 \cdot 3x + 10 \cdot 10$
$= 9x^2 + 30x + 30x + 100$
$= 9x^2 + 60x + 100$

The area is $(9x^2 + 60x + 100)$ sq. meters.

20. $A = l \cdot w$
$= (y^2 + 3y + 7)(y + 5)$
$= y^2(y + 5) + 3y(y + 5) + 7(y + 5)$
$= y^2 \cdot y + y^2 \cdot 5 + 3y \cdot y + 3y \cdot 5$
$\quad + 7 \cdot y + 7 \cdot 5$
$= y^3 + 5y^2 + 3y^2 + 15y + 7y + 35$
$= y^3 + 8y^2 + 22y + 35$

The area is $(y^3 + 8y^2 + 22y + 35)$ sq. inches.

10.4 SOLUTIONS TO EXERCISES

1. $x^9 = x^4 \cdot x^5$
$x^4 = x^4$
$x^7 = x^4 \cdot x^3$

     GCF $= x^4$

2.   The GCF of 2, 8, and 6 is 2.
The GCF of y^2, y^5, and y is y.

     The GCF of $2y^2$, $8y^5$, and $6y$ is $2y$.

3.   The GCF of a^2, a^3, and a is a.
The GCF of b^3, b^2, and b^2 is b^2.

     The GCF of a^2b^3, a^3b^2, and ab^2 is ab^2.

4.   The GCF of 3, 15, and 6 is 3.
The GCF of x^3, x^2, and x^3 is x^2.
The GCF of y^4, y^3, and y^4 is y^3.
The GCF of z^2, z^3, and z^2 is z^2.

     The GCF of $3x^3y^4z^2$, $15x^2y^3z^3$,
and $6x^3y^4z^2$ is $3x^2y^3z^2$.

5. The GCF of 14, 21, and 35 is 7.
The GCF of y, y^2, and y^3 is y.
The GCF of z, z. and z^2 is z.
The GCF of $14yz$, $21y^2z$ and $35y^3z^2$ is $7yz$.

6. The GCF of 18, 27, and 54 is 9.
The GCF of a^4, a^6, and a^7 is a^4.
The GCF of b^5, b^7, and b^5 is b^5.

The GCF of $18a^4b^5$, $27a^6b^7$, and $54a^7b^5$
is $9a^4b^5$.

7. $4x^3 + 8x^2 = 4x^2 \cdot x + 4x^2 \cdot 2$
$= 4x^2(x + 2)$

8. $6y^4 - 3y = 3y \cdot 2y^3 - 3y \cdot 1$
$= 3y(2y^3 - 1)$

9. $20z^4 + 10z^2 = 10z^2 \cdot 2z^2 + 10z^2 \cdot 1$
$= 10z^2(2z^2 + 1)$

10. $12a^3 - 4a^5 = 4a^3 \cdot 3 - 4a^3 \cdot a^2$
$= 4a^3(3 - a^2)$

11. $b^{12} - 2b^8 = b^8 \cdot b^4 - b^8 \cdot 2$
$= b^8(b^4 - 2)$

12. $m^{10} + 5m^4 = m^4 \cdot m^6 + m^4 \cdot 5$
$= m^4(m^6 + 5)$

13. $13x^6 - 26x^4 + 39x^2$
$= 13x^2 \cdot x^4 - 13x^2 \cdot 2x^2 + 13x^2 \cdot 3$
$= 13x^2(x^4 - 2x^2 + 3)$

14. $8y^5 - 4y^3 - 12y^2$
$= 4y^2 \cdot 2y^3 - 4y^2 \cdot y - 4y^2 \cdot 3$
$= 4y^2(2y^3 - y - 3)$

15. $3a^4 - 6b^2 + 9 = 3 \cdot a^4 - 3 \cdot 2b^2 + 3 \cdot 3$
$= 3(a^4 - 2b^2 + 3)$

16. $11m^4 + 22m^3 - 33m$
$= 11m \cdot m^3 + 11m \cdot 2m^2 - 11m \cdot 3$
$= 11m(m^3 + 2m^2 - 3)$

CHAPTER 10 PRACTICE TEST SOLUTIONS

1. $(15x - 7) + (2x + 8) = (15x + 2x) + (-7 + 8)$
$= 17x + 1$

2. $(15x - 7) - (2x + 8) = (15x - 7) + (-2x - 8)$
$= (15x - 2x) + (-7 - 8)$
$= 13x - 15$

3. $(4.9y^2 + 10) + (3.2y^2 - 5y - 17)$
$= (4.9y^2 + 3.2y^2) - 5y + (10 - 17)$
$= 8.1y^2 - 5y - 7$

4. $(4b^2 - 4b - 1) - (6b^2 - 5)$
$= (4b^2 - 4b - 1) + (-6b^2 + 5)$
$= (4b^2 - 6b^2) - 4b + (-1 + 5)$
$= -2b^2 - 4b + 4$

5. $x^2 + 3x - 5 = 6^2 + 3(6) - 5$
$= 36 + 18 - 5$
$= 49$

6. $y^6 \cdot y^8 = y^{6+8}$
$= y^{14}$

7. $(x^5)^9 = x^{5 \cdot 9}$
$= x^{45}$

8. $(3x^6)^3 = 3^3(x^6)^3$
$= 27x^{6 \cdot 3}$
$= 27x^{18}$

9. $(-15y^2)(-4y^8) = (-15)(-4)y^{2+8}$
$= 60y^{10}$

10. $(b^{12})^5(b^2)^3 = b^{12 \cdot 5} \cdot b^{2 \cdot 3}$
$= b^{60} \cdot b^6$
$= b^{60+6}$
$= b^{66}$

11. $(4x^3y)^4(3yx^2)^3 = 4^4(x^3)^4y^4 \cdot 3^3y^3(x^2)^3$
$= 256x^{12}y^4 \cdot 27y^3x^6$
$= 6912x^{18}y^7$

12. $6x(9x^3 - 2.6) = 6x(9x^3) + 6x(-2.6)$
$$= 54x^4 - 15.6x$$

13. $-5a(a^4 - 10a^3 + 1)$
$$= -5a(a^4) - 5a(-10a^3) - 5a(1)$$
$$= -5a^5 + 50a^4 - 5a$$

14. $(x - 6)(x - 7) = x(x - 7) - 6(x - 7)$
$$= x \cdot x + x(-7) - 6 \cdot x - 6(-7)$$
$$= x^2 - 7x - 6x + 42$$
$$= x^2 - 13x + 42$$

15. $(4x + 9)^2 = (4x + 9)(4x + 9)$
$$= 4x(4x + 9) + 9(4x + 9)$$
$$= 4x \cdot 4x + 4x \cdot 9 + 9 \cdot 4x + 9 \cdot 9$$
$$= 16x^2 + 36x + 36x + 81$$
$$= 16x^2 + 72x + 81$$

16. $(y - 7)(y^2 + 7y + 49)$
$$= y(y^2 + 7y + 49) - 7(y^2 + 7y + 49)$$
$$= y \cdot y^2 + y \cdot 7y + y \cdot 49 - 7 \cdot y^2 - 7 \cdot 7y - 7 \cdot 49$$
$$= y^3 + 7y^2 + 49y - 7y^2 - 49y - 343$$
$$= y^3 - 343$$

17. $A = l \cdot w$
$$= (3x + 4)(5x)$$
$$= 3x \cdot 5x + 4 \cdot 5x$$
$$= 15x^2 + 20x$$

$P = 2l + 2w$
$$= 2(3x + 4) + 2(5x)$$
$$= 2 \cdot 3x + 2 \cdot 4 + 2(5x)$$
$$= 6x + 8 + 10x$$
$$= 16x + 8$$

18. $75 = 3 \cdot 5^2$
$90 = 2 \cdot 3^2 \cdot 5$
GCF $= 3 \cdot 5 = 15$

19. The GCF of 4, 18, and 24 is 2.
The GCF of x^6, x^3, and x^9 is x^3.

The GCF of $4x^6$, $18x^3$, $24x^9$ is $2x^3$.

20. $7x^2 - 35 = 7 \cdot x^2 - 7 \cdot 5$
$$= 7(x^2 - 5)$$

21. $y^{10} - 4y^6 = y^6 \cdot y^4 - y^6 \cdot 4$
$$= y^6(y^4 - 4)$$

22. $3x^2 - 6x + 30 = 3 \cdot x^2 - 3 \cdot 2x + 3 \cdot 10$
$$= 3(x^2 - 2x + 10)$$

23. $-16t^2 + 150 = -16(2)^2 + 150$
$$= -16(4) + 150$$
$$= -64 + 150$$
$$= 86$$

The rock's height is 86 feet.

SOLUTIONS TO PRACTICE FINAL EXAMINATION A

1.
$$\begin{array}{r} 62 \\ +\ 57 \\ \hline 119 \end{array}$$

2.
$$\begin{array}{r} {}^{8\ 11} \\ 69\!\!\!/1 \\ -\ 328 \\ \hline 363 \end{array}$$

3.
$$\begin{array}{r} 127 \\ \times\ 13 \\ \hline 381 \\ 127 \\ \hline 1651 \end{array}$$

4.
$$\begin{array}{r} 35 \\ 81\overline{)2835} \\ \underline{243} \\ 405 \\ \underline{405} \\ 0 \end{array}$$

5. $(3^2 - 4) \cdot 8 = (9 - 4) \cdot 8$
$$= 5 \cdot 8$$
$$= 40$$

6. $21 + 12 \div 6 \cdot 4 - 7$
$$= 21 + 2 \cdot 4 - 7$$
$$= 21 + 8 - 7$$
$$= 22$$

7. 2,137,546 rounds to 2,138,000.

8. $4(x^3 - 6) = 4(2^3 - 6)$
$$= 4(8 - 6)$$
$$= 4(2)$$
$$= 8$$

9. Total cost ÷ number   =   cost
              of cans   per can

     $312   ÷   26   = $12

Each can cost $12.

10. $18 - 45 = 18 + (-45)$
$$= -27$$

11. $-22 + 17 = -5$

12. $(-3) \cdot (-30) = 90$

13. $\dfrac{-57}{3} = -19$

14. $(-8)^2 - 24 \div (-6) = 64 - 24 \div (-6)$
$$= 64 - (-4)$$
$$= 64 + 4$$
$$= 68$$

15. $\dfrac{|39 - 57|}{6} = \dfrac{|-18|}{6}$
$$= \dfrac{18}{6}$$
$$= 3$$

16. $\dfrac{8(-11) + 12}{-1(9 - 28)} = \dfrac{-88 + 12}{-1(-19)}$
$$= \dfrac{-76}{19}$$
$$= -4$$

17. $\dfrac{25}{-5} - \dfrac{4^3}{8} = \dfrac{25}{-5} - \dfrac{64}{8}$
$$= -5 - 8$$
$$= -13$$

18. $3x - y = 3(-7) - (-6)$
$$= -21 + 6$$
$$= -15$$

19. $|x - y| + |x| = |-8 - 10| + |-8|$
$$= |-18| + |-8|$$
$$= 18 + 8$$
$$= 26$$

20. $7x - 17 - 12x + 30 = 7x - 12x - 17 + 30$
$$= -5x + 13$$

21. $-10(2x + 9) = -10(2x) - 10(9)$
$$= -20x - 90$$

22. $P = 4s$
$$= 4(17 \text{ feet})$$
$$= 68 \text{ feet}$$

23. $8x + x = 81$
$$9x = 81$$
$$\frac{9x}{9} = \frac{81}{9}$$
$$x = 9$$

24. $22 = 3x - 14x$
$$22 = -11x$$
$$\frac{22}{-11} = \frac{-11x}{-11}$$
$$-2 = x$$

25. $19x + 12 - 18x - 30 = 20$
$$x - 18 = 20$$
$$x - 18 + 18 = 20 + 18$$
$$x = 38$$

26. $5(x + 4) = 0$
$$5x + 20 = 0$$
$$5x + 20 - 20 = 0 - 20$$
$$5x = -20$$
$$\frac{5x}{5} = \frac{-20}{5}$$
$$x = -4$$

27. $10 + 6(2y - 1) = 28$
$$10 + 12y - 6 = 28$$
$$12y + 4 = 28$$
$$12y + 4 - 4 = 28 - 4$$
$$12y = 24$$
$$\frac{12y}{12} = \frac{24}{12}$$
$$y = 2$$

28. $4(3x - 5) = 2(7x + 8)$
$$12x - 20 = 14x + 16$$
$$12x - 20 + 20 = 14x + 16 + 20$$
$$12x = 14x + 36$$
$$12x - 14x = 14x + 36 - 14x$$
$$-2x = 36$$
$$\frac{-2x}{-2} = \frac{36}{-2}$$
$$x = -18$$

29. Let x = the unknown number
$$3x + 7x = 90$$
$$10x = 90$$
$$\frac{10x}{10} = \frac{90}{10}$$
$$x = 9$$

The number is 9.

30. Let x = length
$$P = 4s$$
$$72 = 4x$$
$$\frac{72}{4} = \frac{4x}{4}$$
$$18 = x$$

The side length is 18 feet.

31. $\dfrac{4}{5} \cdot \dfrac{35}{8} = \dfrac{2 \cdot 2 \cdot 5 \cdot 7}{5 \cdot 2 \cdot 2 \cdot 2}$
$$= \frac{7}{2}$$

32. $\dfrac{5x}{7} - \dfrac{3x}{7} = \dfrac{5x - 3x}{7}$
$$= \frac{2x}{7}$$

33. $\dfrac{xy^3}{z^2} \cdot \dfrac{z^3}{x^4y} = \dfrac{x \cdot y \cdot y \cdot y \cdot z \cdot z \cdot z}{z \cdot z \cdot x \cdot x \cdot x \cdot x \cdot y}$

$= \dfrac{y \cdot y \cdot z}{x \cdot x \cdot x}$

$= \dfrac{y^2z}{x^3}$

34. $\dfrac{6y}{13} + \dfrac{5}{26} = \dfrac{6y \cdot 2}{13 \cdot 2} + \dfrac{5}{26}$

$= \dfrac{12y}{26} + \dfrac{5}{26}$

$= \dfrac{12y + 5}{26}$

35. $4\dfrac{1}{5} \div \dfrac{7}{25} = \dfrac{21}{5} \div \dfrac{7}{25}$

$= \dfrac{21}{5} \cdot \dfrac{25}{7}$

$= \dfrac{3 \cdot 7 \cdot 5 \cdot 5}{5 \cdot 7}$

$= 15$

36. $\begin{array}{r} 4\dfrac{1}{3} = 4\dfrac{6}{18} \\[2mm] 3\dfrac{5}{6} = 3\dfrac{15}{18} \\[2mm] + 8\dfrac{2}{9} = 8\dfrac{4}{18} \\ \hline 15\dfrac{25}{18} \end{array}$

$= 15 + 1\dfrac{7}{18}$

$= 16\dfrac{7}{18}$

37. $18 \div 3\dfrac{4}{9} = 18 \div \dfrac{31}{9}$

$= \dfrac{18}{1} \cdot \dfrac{9}{31}$

$= \dfrac{162}{31}$

or $5\dfrac{7}{31}$

38. $20y^2 \div \dfrac{y}{5} = \dfrac{20y^2}{1} \cdot \dfrac{5}{y}$

$= \dfrac{20y \cdot y \cdot 5}{y}$

$= 100y$

39. $-\dfrac{8}{7} \div \dfrac{64}{21} = -\dfrac{8}{7} \cdot \dfrac{21}{64}$

$= -\dfrac{8 \cdot 3 \cdot 7}{7 \cdot 8 \cdot 8}$

$= -\dfrac{3}{8}$

40. $\left(\dfrac{15}{8} \cdot \dfrac{32}{3} \right) \div 6 = \left(\dfrac{3 \cdot 5 \cdot 8 \cdot 4}{8 \cdot 3} \right) \div 6$

$= 20 \div 6$

$= \dfrac{20}{6} = \dfrac{10}{3}$

41. $\dfrac{6 + \dfrac{3}{5}}{4 - \dfrac{7}{10}} = \dfrac{6 + \dfrac{3}{5}}{4 - \dfrac{7}{10}} \cdot \dfrac{10}{10}$

$= \dfrac{6(10) + \dfrac{3}{5} \cdot \dfrac{10}{1}}{4(10) - \dfrac{7}{10} \cdot \dfrac{10}{1}}$

$= \dfrac{60 + 6}{40 - 7}$

$= \dfrac{66}{33}$

$= 2$

42. $$\frac{x}{3} + x = -\frac{16}{9}$$

$$9\left(\frac{x}{3} + x\right) = 9\left(-\frac{16}{9}\right)$$

$$9\left(\frac{x}{3}\right) + 9(x) = 9\left(-\frac{16}{9}\right)$$

$$3x + 9x = -16$$

$$12x = -16$$

$$\frac{12x}{12} = \frac{-16}{12}$$

$$x = -\frac{4}{3}$$

43. $$\frac{5}{9} + \frac{x}{6} = \frac{2}{3} - \frac{x}{12}$$

$$36\left(\frac{5}{9} + \frac{x}{6}\right) = 36\left(\frac{2}{3} - \frac{x}{12}\right)$$

$$\frac{36}{1} \cdot \frac{5}{9} + \frac{36}{1} \cdot \frac{x}{6} = \frac{36}{1} \cdot \frac{2}{3} - \frac{36}{1} \cdot \frac{x}{12}$$

$$20 + 6x = 24 - 3x$$
$$20 + 6x + 3x = 24 - 3x + 3x$$
$$20 + 9x = 24$$
$$20 + 9x - 20 = 24 - 20$$
$$9x = 4$$

$$\frac{9x}{9} = \frac{4}{9}$$

$$x = \frac{4}{9}$$

44. $$xy = \left(-\frac{2}{5}\right)\left(2\frac{3}{10}\right)$$

$$= -\frac{2}{5} \cdot \frac{23}{10}$$

$$= -\frac{2 \cdot 23}{5 \cdot 2 \cdot 5}$$

$$= -\frac{23}{25}$$

45.
```
 1 2   1
11.654
 3.71
+ 9.827
25.191
```

46.
```
    1
 51.6
+ 25.82
 77.42
```

$$-51.6 - 25.82 = -77.42$$

47.
```
   14.6
 × 3.25
   730
   292
   438
47.450
```

48. $(-1.2)^2 + 3.7 = 1.44 + 3.7$
$$= 5.14$$

49. $$\frac{0.11 - 3.75}{0.2} = \frac{-3.64}{0.2}$$

$$= -18.2$$

50. 126.6457 rounds to 126.65

51. $$0.72 = \frac{72}{100}$$

$$= \frac{18}{25}$$

52.
```
      0.045
200)9.000
    8 00
    1 000
    1 000
        0
```

$$\frac{9}{200} = 0.045$$

53. $-2\sqrt{36} + \sqrt{25} = -2(6) + 5$
$$= -12 + 5$$
$$= -7$$

54. $V = s^3$

$$= \left(2\frac{5}{6}\right)^3$$

$$= \left(\frac{17}{6}\right)^3$$

$$= \frac{4913}{216}$$

or $22\frac{161}{216}$

The volume is $22\frac{161}{216}$ cu. centimeters.

55. $A = l \cdot w$
$$= (9 \text{ feet}) \cdot (2.7 \text{ feet})$$
$$= 24.3 \text{ sq. feet}$$

The area is 24.3 sq. feet.

56. $-2x + 3y = 12$

Let $x = 0$: $-2(0) + 3y = 12$
$$3y = 12$$

$$\frac{3y}{3} = \frac{12}{3}$$

$$y = 4$$
$$(0, 4)$$

Let $y = 0$: $-2x + 3(0) = 12$
$$-2x = 12$$

$$\frac{-2x}{-2} = \frac{12}{-2}$$

$$x = -6$$
$$(-6, 0)$$

Let $y = 2$: $-2x + 3(2) = 12$
$$-2x + 6 = 12$$
$$-2x + 6 - 6 = 12 - 6$$
$$-2x = 6$$

$$\frac{-2x}{-2} = \frac{6}{-2}$$

$$x = -3$$
$$(-3, 2)$$

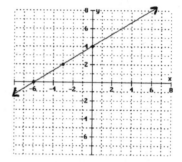

57. $y = 5x + 1$

Let $x = 0$: $y = 5(0) + 1$
$$= 1$$
$$(0, 1)$$

Let $x = 1$: $y = 5(1) + 1$
$$= 5 + 1$$
$$= 6$$
$$(1, 6)$$

Let $y = -4$: $-4 = 5x + 1$
$$-4 - 1 = 5x + 1 - 1$$
$$-5 = 5x$$

$$\frac{-5}{5} = \frac{5x}{5}$$

$$-1 = x$$
$$(-1, -4)$$

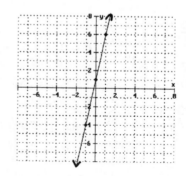

58. $x + y = 3$

Let $x = 0$: $0 + y = 3$
$y = 3$
$(0, 3)$

Let $y = 0$: $x + 0 = 3$
$x = 3$
$x = 3$
$(3, 0)$

Let $x = 1$: $1 + y = 3$
$1 + y - 1 = 3 - 1$
$y = 2$
$(1, 2)$

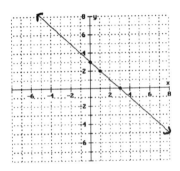

59. $x - 5 = 0$
$x - 5 + 5 = 0 + 5$
$x = 5$

This is a vertical line that crosses the x-axis at 5.

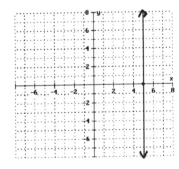

60. $y + 4 = 0$
$y + 4 - 4 = 0 - 4$
$y = -4$

This is a horizontal line that crosses the y-axis at -4.

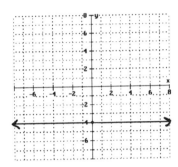

61. $4x - 5y = 20$

Let $x = 0$: $4(0) - 5y = 20$
$-5y = 20$

$$\frac{-5y}{-5} = \frac{20}{-5}$$

$y = -4$
$(0, -4)$

Let $y = 0$: $4x - 5(0) = 20$
$4x = 20$

$$\frac{4x}{4} = \frac{20}{4}$$

$x = 5$
$(5, 0)$

Let $x = 2$: $4(2) - 5y = 20$
$8 - 5y = 20$
$8 - 5y - 8 = 20 - 8$
$-5y = 12$

$$\frac{-5y}{-5} = \frac{12}{-5}$$

$$y = -\frac{12}{5}$$

$$\left(2, -\frac{12}{5}\right)$$

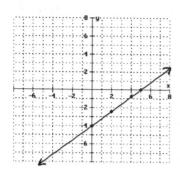

62. mean $= \dfrac{58 + 42 + 67 + 71 + 49 + 52 + 60}{7}$

$= \dfrac{399}{7}$

$= 57$

63. 16, 16, 21, 28, <u>37</u>, 47, 75, 83, 92
median $= 37$

64. 16, 16, 16, 17, 30, 31, 31, 45, 80, 81, 90
mode $= 16$

65.

Grade	Point Value of Grade	Credit Hours	(Point value) · (Credit Hours)
A	4	3	12
A	4	4	16
B	3	5	15
C	2	3	6
	Totals	15	49

grade point average $= \dfrac{49}{15}$

$= 3.27$

66. $\dfrac{365 \text{ bushes}}{657 \text{ bushes}} = \dfrac{365}{657}$

$= \dfrac{5}{9}$

67. $\dfrac{972 \text{ miles}}{18 \text{ hours}} = 54$ miles/hour

68. $\dfrac{15}{x} = \dfrac{105}{98}$

$15 \cdot 98 = x \cdot 105$

$\dfrac{1470}{105} = \dfrac{105x}{105}$

$14 = x$

69. $\dfrac{\frac{9}{5}}{\frac{3}{4}} = \dfrac{x}{\frac{25}{18}}$

$\dfrac{9}{5} \cdot \dfrac{25}{18} = \dfrac{3}{4} \cdot x$

$\dfrac{5}{2} = \dfrac{3}{4}x$

$\dfrac{4}{3} \cdot \dfrac{5}{2} = \dfrac{4}{3} \cdot \dfrac{3}{4}x$

$\dfrac{10}{3} = x$

70.
$$\begin{array}{r} 45 \text{ ft } 6 \text{ in.} \\ 12\overline{)546} \\ \underline{48} \\ 66 \\ \underline{60} \\ 6 \end{array}$$

546 in. $= 45$ ft 6 in.

71. $7\frac{1}{2}$ gal $= \frac{15}{2}$ gal $\cdot \frac{4 \text{ qt}}{1 \text{ gal}}$

$= \frac{15}{2} \cdot 4$ qt

$= 30$ qt

72. g   dg   cg   mg

           ↑

3 units right

62 g = 62.000

       ↑

    = 62,000 mg

73. L   dl   cl   ml

↑

3 units left

5700 ml = 5700.

       ↑

    = 5.7 L

74. $C = \frac{5(F - 32)}{9}$

$= \frac{5(95 - 32)}{9}$

$= \frac{5(63)}{9}$

$= 35$

35°C

75.   12 lb   6 oz

 + 10 lb 14 oz

   22 lb 20 oz

= 22 lb 1 lb 4 oz

= 23 lb 4 oz

76. 2.7 km = 2700 m

    2700 m

 +   231 m

    2931 m

231 m = 0.231 km

    2.7   km

+ 0.231 km

    2.931 km

77. 9 cm = 90 mm

   90 mm

− 15 mm

   75 mm

15 mm = 1.5 cm

   9.0 cm

− 1.5 cm

   7.5 cm

78. 0.0012 = 0.0012

        ↑

     = 0.12%

79. 32.8% = 32.8%

      ↑

     = 0.328

80. $\frac{1}{20} = \frac{1}{20} \cdot 100\%$

$= \frac{100}{20}\%$

$= 5\%$

81. $x = 68\% \cdot 900$
$x = 0.68 \cdot 900$
$x = 612$

82. $182 = 35\% \cdot x$
$182 = 0.35x$

$\frac{182}{0.35} = \frac{0.35x}{0.35}$

$520 = x$

83. $40.32 = x \cdot 96$

$\frac{40.32}{96} = \frac{x \cdot 96}{96}$

$0.42 = x$
$42\% = x$

84. What is 22% of 210?

$x = 22\% \cdot 210$
$x = 0.22 \cdot 210$
$x = 46.2$

There are 46.2 lb of sunflower seeds in the mixture.

85. Total amount $= \$375 + 6.5\% \cdot \$375$
$\phantom{\text{Total amount }} = \$375 + 0.065 \cdot \$375$
$\phantom{\text{Total amount }} = \$375 + \$24.38$
$\phantom{\text{Total amount }} = \$399.38$

The total amount charged is $399.38.

86. quarterly
  rate = 6%
  time = 5 years

compound interest factor = 1.34686
(See Appendix F in textbook)

Total amount = original principal · compound interest factor
$\phantom{\text{Total amount }} = \$9600(1.34686)$
$\phantom{\text{Total amount }} = \$12,929.86$

87. commission = commission rate · sales
$\phantom{\text{commission }} = 2.6\% \cdot 18,700$
$\phantom{\text{commission }} = (0.026)(18,700)$
$\phantom{\text{commission }} = 486.20$

Her commission was $486.20.

88. $(9y + 15) + (-6y + 8) = (9y - 6y) + (15 + 8)$
$ = 3y + 23$

89. $(12x - 5) - (x - 1) = (12x - 5) + (-x + 1)$
$ = (12x - x) + (-5 + 1)$
$ = 11x - 4$

90. $x^2 + x - 7 = 5^2 + 5 - 7$
$ = 25 + 5 - 7$
$ = 23$

91. $y^{15} \cdot y^{16} = y^{15+16}$
$\phantom{y^{15} \cdot y^{16} } = y^{31}$

92. $(x^8)^9 = x^{8 \cdot 9}$
$ = x^{72}$

93. $(a^3)^5(a^{10})^6 = a^{3 \cdot 5} \cdot a^{10 \cdot 6}$
$\phantom{(a^3)^5(a^{10})^6 } = a^{15} \cdot a^{60}$
$\phantom{(a^3)^5(a^{10})^6 } = a^{15+60}$
$\phantom{(a^3)^5(a^{10})^6 } = a^{75}$

94. $-9y(y^2 - 4y + 3)$
$= -9y \cdot y^2 - 9y(-4y) - 9y \cdot 3$
$= -9y^3 + 36y^2 - 27y$

95. $(x + 11)(x - 9)$
$= x(x - 9) + 11(x - 9)$
$= x \cdot x + x \cdot (-9) + 11 \cdot x + 11 \cdot (-9)$
$= x^2 - 9x + 11x - 99$
$= x^2 + 2x - 99$

96. $(4x + 13)^2$
$= (4x + 13)(4x + 13)$
$= 4x(4x + 13) + 13(4x + 13)$
$= 4x \cdot 4x + 4x \cdot 13 + 13 \cdot 4x + 13 \cdot 13$
$= 16x^2 + 52x + 52x + 169$
$= 16x^2 + 104x + 169$

97. $(b - 8)(b^2 + 8b + 64)$
$= b(b^2 + 8b + 64) - 8(b^2 + 8b + 64)$
$= b \cdot b^2 + b \cdot 8b + b \cdot 64 - 8 \cdot b^2$
$ - 8 \cdot 8b - 8 \cdot 64$
$= b^3 + 8b^2 + 64b - 8b^2 - 64b - 512$
$= b^3 - 512$

98. The GCF of 12, 16, and 28 is 4.
The GCF of x^7, x^2, and x^5 is x^2.

The GCF of $12x^7$, $16x^2$, $28x^5$ is $4x^2$.

99. $8x^3 - 64x^5 = 8x^3 \cdot 1 - 8x^3 \cdot 8x^2$
$ = 8x^3(1 - 8x^2)$

100. $10y^2 - 20y = 10y \cdot y - 10y \cdot 2$
$ = 10y(y - 2)$

SOLUTIONS TO PRACTICE FINAL EXAMINATION B

1.
$$\begin{array}{r} \overset{1}{} \\ 76 \\ +\ 89 \\ \hline 165 \end{array}$$

2.
$$\begin{array}{r} {}^{12} \\ 6\,\cancel{7}\,{}^{12} \\ 732 \\ -\ 468 \\ \hline 264 \end{array}$$

3.
$$\begin{array}{r} 212 \\ \times\ 18 \\ \hline 1696 \\ 212 \\ \hline 3816 \end{array}$$

4.
$$\begin{array}{r} 91 \\ 65\overline{)5915} \\ \underline{585} \\ 65 \\ \underline{65} \\ 0 \end{array}$$

5. $(2^3 - 3) \cdot 7 = (8 - 3) \cdot 7$
   $= 5 \cdot 7$
   $= 35$

6. $13 + 18 \div 9 \cdot 5 - 10 = 13 + 2 \cdot 5 - 10$
   $= 13 + 10 - 10$
   $= 23 - 10$
   $= 13$

7. 561,382 rounds to 561,400.

8. $9(x^2 - 12) = 9(7^2 - 12)$
   $= 9(49 - 12)$
   $= 9(37)$
   $= 333$

9.
Total cost	÷	number of cases	=	cost per case
$76.86	÷	14	=	$5.49

10. $21 - 58 = 21 + (-58)$
    $= -37$

11. $-37 + 10 = -27$

12. $(-10) \cdot 8 = -80$

13. $\dfrac{-91}{-7} = 13$

14. $(-9)^2 - 35 \div (-5) = 81 - 35 \div (-5)$
    $= 81 - (-7)$
    $= 81 + 7$
    $= 88$

15. $\dfrac{|9 - 60|}{-17} = \dfrac{|-51|}{-17}$

    $= \dfrac{51}{-17}$

    $= -3$

16. $\dfrac{4(-12) + 8}{-1(-2 - 6)} = \dfrac{-48 + 8}{-1(-8)}$

    $= \dfrac{-40}{8}$

    $= -5$

17. $\dfrac{30}{-6} - \dfrac{3^3}{9} = \dfrac{30}{-6} - \dfrac{27}{9}$

    $= -5 - 3$
    $= -8$

18. $5x - y = 5(-9) - (-4)$
    $= -45 + 4$
    $= -41$

19. $|x + y| - |x| = |-5 + 12| - |-5|$
    $= |7| - |-5|$
    $= 7 - 5$
    $= 2$

20. $14x + 21 - 8x - 49 = 14x - 8x + 21 - 49$
$$= 6x - 28$$

21. $-30(2x + 8) = -30(2x) - 30(8)$
$$= -60x - 240$$

22. $P = 4s$
$$= 4(19 \text{ feet})$$
$$= 76 \text{ feet}$$

23. $11x + 3x = 28$
$$14x = 28$$
$$\frac{14x}{14} = \frac{28}{14}$$
$$x = 2$$

24. $32 = 7x - 15x$
$$32 = -8x$$
$$\frac{32}{-8} = \frac{-8x}{-8}$$
$$-4 = x$$

25. $36x + 11 - 35x - 50 = 10$
$$x - 39 = 10$$
$$x - 39 + 39 = 10 + 39$$
$$x = 49$$

26. $6(x - 3) = 0$
$$6x - 18 = 0$$
$$6x - 18 + 18 = 0 + 18$$
$$6x = 18$$
$$\frac{6x}{6} = \frac{18}{6}$$
$$x = 3$$

27. $9 + 5(3y - 2) = 29$
$$9 + 15y - 10 = 29$$
$$15y - 1 = 29$$
$$15y - 1 + 1 = 29 + 1$$
$$15y = 30$$
$$\frac{15y}{15} = \frac{30}{15}$$
$$y = 2$$

28. $8(2x - 6) = 4(3x + 5)$
$$16x - 48 = 12x + 20$$
$$16x - 48 + 48 = 12x + 20 + 48$$
$$16x = 12x + 68$$
$$16x - 12x = 12x + 68 - 12x$$
$$4x = 68$$
$$\frac{4x}{4} = \frac{68}{4}$$
$$x = 17$$

29. Let x = the unknown number
$$5x + 8x = 39$$
$$13x = 39$$
$$\frac{13x}{13} = \frac{39}{13}$$
$$x = 3$$

The number is 3.

30. Let x = length
$$P = 4s$$
$$84 = 4x$$
$$\frac{84}{4} = \frac{4x}{4}$$
$$21 = x$$

The side length is 21 feet.

31. $\dfrac{7}{9} \cdot \dfrac{54}{49} = \dfrac{7 \cdot 9 \cdot 6}{9 \cdot 7 \cdot 7}$
$$= \frac{6}{7}$$

32. $\dfrac{7x}{5} - \dfrac{2x}{5} = \dfrac{7x - 2x}{5}$
$$= \frac{5x}{5}$$
$$= x$$

33. $\dfrac{x^2y}{z} \cdot \dfrac{z^2}{x^3y^2} = \dfrac{x \cdot x \cdot y \cdot z \cdot z}{z \cdot x \cdot x \cdot x \cdot y \cdot y}$

$= \dfrac{z}{xy}$

34. $\dfrac{8y}{17} + \dfrac{4}{51} = \dfrac{18y \cdot 3}{17 \cdot 3} + \dfrac{4}{51}$

$= \dfrac{54y}{51} + \dfrac{4}{51}$

$= \dfrac{54y + 4}{51}$

35. $2\dfrac{1}{8} \div \dfrac{3}{4} = \dfrac{17}{8} \div \dfrac{3}{4}$

$= \dfrac{17}{8} \cdot \dfrac{4}{3}$

$= \dfrac{17 \cdot 4}{4 \cdot 2 \cdot 3}$

$= \dfrac{17}{6}$

36. $7\dfrac{1}{2} = 7\dfrac{12}{24}$

$4\dfrac{3}{8} = 4\dfrac{9}{24}$

$+ 9\dfrac{5}{6} = 9\dfrac{20}{24}$

$\rule{3cm}{0.4pt}$

$20\dfrac{41}{24}$

$= 20 + 1\dfrac{17}{24}$

$= 21\dfrac{17}{24}$

37. $21 \div 6\dfrac{3}{7} = \dfrac{21}{1} \div \dfrac{45}{7}$

$= \dfrac{21}{1} \cdot \dfrac{7}{45}$

$= \dfrac{3 \cdot 7 \cdot 7}{1 \cdot 3 \cdot 3 \cdot 5}$

$= \dfrac{49}{15}$

38. $10y^3 \div \dfrac{y}{4} = \dfrac{10y^3}{1} \cdot \dfrac{4}{y}$

$= \dfrac{10y \cdot y \cdot y \cdot 4}{1 \cdot y}$

$= 40y^2$

39. $-\dfrac{6}{7} \div \left(-\dfrac{36}{35}\right) = -\dfrac{6}{7} \cdot \left(-\dfrac{35}{36}\right)$

$= \dfrac{6 \cdot 5 \cdot 7}{7 \cdot 6 \cdot 6}$

$= \dfrac{5}{6}$

40. $4 \div \left(\dfrac{2}{3} \cdot \dfrac{9}{16}\right) = \dfrac{4}{1} \div \left(\dfrac{2 \cdot 3 \cdot 3}{3 \cdot 2 \cdot 2 \cdot 2 \cdot 2}\right)$

$= \dfrac{4}{1} \div \dfrac{3}{8}$

$= \dfrac{4}{1} \cdot \dfrac{8}{3}$

$= \dfrac{32}{3}$

41. $\dfrac{\dfrac{2}{3} + \dfrac{3}{4}}{\dfrac{1}{6} - \dfrac{1}{12}} = \dfrac{\dfrac{2}{3} + \dfrac{3}{4}}{\dfrac{1}{6} - \dfrac{1}{12}} \cdot \dfrac{12}{12}$

$= \dfrac{\dfrac{12}{1} \cdot \dfrac{2}{3} + \dfrac{12}{1} \cdot \dfrac{3}{4}}{\dfrac{12}{1} \cdot \dfrac{1}{6} - \dfrac{12}{1} \cdot \dfrac{1}{12}}$

$= \dfrac{8 + 9}{2 - 1}$

$= \dfrac{17}{1}$

$= 17$

42.

$$\frac{x}{6} + x = -\frac{5}{12}$$

$$12\left(\frac{x}{6} + x\right) = 12\left(-\frac{5}{12}\right)$$

$$\frac{12}{1} \cdot \frac{x}{6} + 12 \cdot x = \frac{12}{1} \cdot \left(-\frac{5}{12}\right)$$

$$2x + 12x = -5$$
$$14x = -5$$
$$\frac{14x}{14} = \frac{-5}{14}$$
$$x = -\frac{5}{14}$$

43.

$$\frac{3}{8} + \frac{x}{16} = \frac{1}{4} - \frac{x}{2}$$

$$16\left(\frac{3}{8} + \frac{x}{16}\right) = 16\left(\frac{1}{4} - \frac{x}{2}\right)$$

$$\frac{16}{1} \cdot \frac{3}{8} + \frac{16}{1} \cdot \frac{x}{16} = \frac{16}{1} \cdot \frac{1}{4} - \frac{16}{1} \cdot \frac{x}{2}$$

$$6 + x = 4 - 8x$$
$$6 + x + 8x = 4 - 8x + 8x$$
$$6 + 9x = 4$$
$$6 + 9x - 6 = 4 - 6$$
$$9x = -2$$
$$\frac{9x}{9} = \frac{-2}{9}$$
$$x = -\frac{2}{9}$$

44.

$$\frac{x}{y} = \frac{-\frac{3}{7}}{1\frac{1}{9}}$$

$$= \frac{-\frac{3}{7}}{\frac{10}{9}}$$

$$= \frac{-\frac{3}{7}}{\frac{10}{9}} \cdot \frac{63}{63}$$

$$= \frac{-27}{70}$$

45.

$$\overset{1}{}$$
$$\begin{array}{r} 8.23 \\ 5.607 \\ + 1.9 \\ \hline 15.737 \end{array}$$

46.

$$\overset{1\,1}{}$$
$$\begin{array}{r} 18.6 \\ \underline{31.5} \\ 50.1 \end{array}$$

$$-18.6 - 31.5 = -50.1$$

47.

$$\begin{array}{r} 20.38 \\ \times\ \ 2.7 \\ \hline 14266 \\ \underline{4076} \\ 55.026 \end{array}$$

48. $(-2.3)^2 + 4.1 = 5.29 + 4.1$
$$= 9.39$$

49. $\dfrac{0.6 - 2.34}{-0.3} = \dfrac{-1.74}{-0.3}$

$$= 5.8$$

50. 29.05$\underline{7}$2 rounds to 29.057

51. $0.68 = \dfrac{68}{100}$

$$= \dfrac{17}{25}$$

52.

$$\begin{array}{r} 0.214 \\ 500\overline{)107.000} \\ \underline{100\,0} \\ 700 \\ \underline{500} \\ 2000 \\ \underline{2000} \\ 0 \end{array}$$

$$\frac{107}{500} = 0.214$$

53. $-7\sqrt{64} - \sqrt{100} = -7(8) - 10$
$$= -56 - 10$$
$$= -66$$

54. $V = s^3$
$$= \left(1\frac{3}{7}\right)^3$$
$$= \left(\frac{10}{7}\right)^3$$
$$= \frac{1000}{343}$$
$$\text{or } 2\frac{314}{343}$$

The volume is $2\frac{314}{343}$ cubic inches.

55. $A = l \cdot w$
$$= (2.3 \text{ feet}) \cdot (4.9 \text{ feet})$$
$$= 11.27 \text{ sq. feet}$$

The area is 11.27 sq. feet.

56. $4x - 7y = 28$

Let $x = 0$: $4(0) - 7y = 28$
$$-7y = 28$$
$$\frac{-7y}{-7} = \frac{28}{-7}$$
$$y = -4$$
$$(0, -4)$$

Let $y = 0$: $4x - 7(0) = 28$
$$4x = 28$$
$$\frac{4x}{4} = \frac{28}{4}$$
$$x = 7$$
$$(7, 0)$$

Let $x = -1$: $4(-1) - 7y = 28$
$$-4 - 7y = 28$$
$$-4 - 7y + 4 = 28 + 4$$
$$-7y = 32$$

$$\frac{-7y}{-7} = \frac{32}{-7}$$
$$y = -\frac{32}{7} \quad \text{or} \quad -4\frac{4}{7}$$
$$\left(-1, -4\frac{4}{7}\right)$$

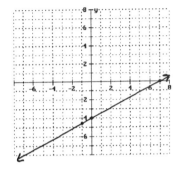

57. $y = 2x - 3$
Let $x = 0$: $y = 2(0) - 3$
$$= -3$$
$$(0, -3)$$

Let $y = 0$: $0 = 2x - 3$
$$0 + 3 = 2x - 3 + 3$$
$$3 = 2x$$
$$\frac{3}{2} = \frac{2x}{2}$$
$$\frac{3}{2} = x$$
$$\left(\frac{3}{2}, 0\right)$$

Let $x = 2$: $y = 2(2) - 3$
$$= 4 - 3$$
$$= 1$$
$$(2, 1)$$

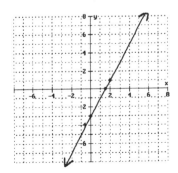

58. $x + 2y = 4$

Let $x = 0$: $0 + 2y = 4$
$$2y = 4$$
$$\frac{2y}{2} = \frac{4}{2}$$
$$y = 2$$
$$(0, 2)$$

Let $y = 0$: $x + 2(0) = 4$
$$x = 4$$
$$(4, 0)$$

Let $x = 2$: $2 + 2y = 4$
$$2 + 2y - 2 = 4 - 2$$
$$2y = 2$$
$$\frac{2y}{2} = \frac{2}{2}$$
$$y = 1$$
$$(2, 1)$$

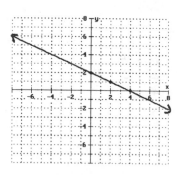

59. $x + 4 = 0$
$$x + 4 + (-4) = 0 + (-4)$$
$$x = -4$$

This is a vertical line that crosses the
x-axis at -4.

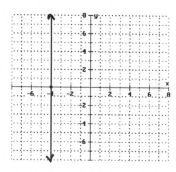

60. $2y - 3 = 0$
$$2y - 3 + 3 = 0 + 3$$
$$2y = 3$$
$$\frac{2y}{2} = \frac{3}{2}$$
$$y = \frac{3}{2}$$

This is a horizontal line that crosses the
y-axis at $\frac{3}{2}$.

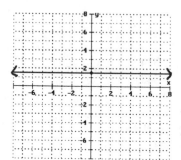

61. $5x + 6y = -30$

Let $x = 0$: $5(0) + 6y = -30$
$$6y = -30$$
$$\frac{6y}{6} = \frac{-30}{6}$$
$$y = -5$$
$$(0, -5)$$

Let $y = 0$: $5x + 6(0) = -30$
$$5x = -30$$
$$\frac{5x}{5} = \frac{-30}{5}$$
$$x = -6$$
$$(-6, 0)$$

Let $x = 2$: $5(2) + 6y = -30$
$$10 + 6y = -30$$
$$10 + 6y - 10 = -30 - 10$$
$$6y = -40$$

$$\frac{6y}{6} = \frac{-40}{6}$$

$$y = -\frac{20}{3} \quad \text{or} \quad -6\frac{2}{3}$$

$$\left(2,\ -6\frac{2}{3}\right)$$

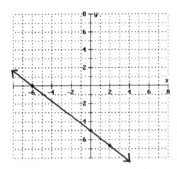

62. $\text{mean} = \dfrac{108 + 112 + 137 + 142 + 98 + 156 + 110 + 118}{8}$

      $= \dfrac{981}{8}$

      $= 122.625$

63. 12, 14, 16, 17, 25, <u>32</u>, <u>36</u>, 49, 57, 58, 61, 80

      $\text{median} = \dfrac{32 + 36}{2}$

              $= \dfrac{68}{2}$

              $= 34$

64. 10, 10, 10, 28, 29, 37, 37, 49, 51, 52, 52, 52, 52, 91

      mode = 52

65.

Grade	Point Value of Grade	Credit Hours	(Point value) · (Credit Hours)
B	3	3	9
A	4	3	12
C	2	4	8
C	2	3	6
B	3	5	15
	Totals	18	50

grade point average $= \dfrac{50}{18}$

$= 2.78$

66. $\dfrac{62 \text{ marbles}}{248 \text{ marbles}} = \dfrac{62}{248}$

$= \dfrac{1}{4}$

67. $\dfrac{550 \text{ miles}}{25 \text{ gal}} = 22 \text{ miles/gal}$

68. $\dfrac{9}{60} = \dfrac{x}{20}$

$9 \cdot 20 = 60 \cdot x$
$180 = 60x$

$\dfrac{180}{60} = \dfrac{60x}{60}$
$3 = x$

70. $1092 \text{ in.} = 1092 \text{ in.} \cdot \dfrac{1 \text{ ft}}{12 \text{ in.}}$

$= \dfrac{1092}{12} \text{ ft}$

$= 91 \text{ ft}$

71. $58 \text{ qt} = 58 \text{ qt} \cdot \dfrac{1 \text{ gal}}{4 \text{ qt}}$

$= \dfrac{58}{4} \text{ gal}$

$= 14\dfrac{1}{2} \text{ gal}$

72. g dg cg mg

3 units left

$12000 \text{ mg} = 12000$

$= 12 \text{ g}$

73. L dl cl ml

3 units right

$16 \text{ L} = 16.000$

$= 16000 \text{ ml}$

74. $F = \dfrac{9C}{5} + 32$

$= \dfrac{9(20)}{5} + 32$

$= 36 + 32$
$= 68$

$68°F$

75. 19 lb 3 oz = 18 lb 19 oz
$\underline{-5 \text{ lb}\ \ 12 \text{ oz} = -5 \text{ lb}\ \ 12 \text{ oz}}$
13 lb 7 oz

76. 3.8 kg = 3800 g

$$\begin{array}{r} 3800 \text{ g} \\ + \ 160 \text{ g} \\ \hline 3960 \text{ g} \end{array}$$

or   160 g = 0.16 kg

$$\begin{array}{r} 3.8 \ \text{ kg} \\ + \ 0.16 \text{ kg} \\ \hline 3.96 \text{ kg} \end{array}$$

77.
$$\begin{array}{r} 2 \text{ ft} \ \ 3 \text{ in.} \\ \times \ \ \ \ \ \ \ 8 \\ \hline 16 \text{ ft} \ \ 24 \text{ in.} \end{array}$$
= 16 ft   2 ft
= 18 ft

78. $0.126 = 0.12\underset{\uparrow}{6}$
       = 12.6%

79. $0.21\% = 00.21\underset{\uparrow}{\%}$
      = 0.0021

80. $\dfrac{6}{75} = \dfrac{6}{75} \cdot 100\%$

    $= \dfrac{600}{75}\%$

    $= 8\%$

81. $x = 5.4\% \cdot 1200$
   $x = 0.054 \cdot 1200$
   $x = 64.8$

82. $172.2 = x \cdot 820$

   $\dfrac{172.2}{820} = \dfrac{820x}{820}$

     $0.21 = x$
     $21\% = x$

83. $28.5 = 30\% \cdot x$
   $28.5 = 0.30x$

   $\dfrac{28.5}{0.30} = \dfrac{0.30x}{0.30}$

    $95 = x$

84. $I = P \cdot R \cdot T$

    $= \$1600 \cdot 8.75\% \cdot 2\dfrac{1}{2}$

    $= \$1600 \cdot 0.0875 \cdot \dfrac{5}{2}$

    $= \$350$

85.   Total = \$7000 + \$1248
          = \$8248

    4 years = 4 · 12 months = 48 months

    monthly payment $= \dfrac{\$8248}{48}$

               = \$171.83

86. commission = commission rate · sales
              = 4.5% · 283,000
              = (0.045)(283,000)
              = 12,735

Her commission was \$12,735.

87. $(8x^2 - 6x + 1) + (5x^2 - 4x - 10)$
   $= (8x^2 + 5x^2) + (-6x - 4x) + (1 - 10)$
   $= 13x^2 - 10x - 9$

88. $(9x^3 - x + 5) - (4x^2 - 7x + 6)$
   $= (9x^3 - x + 5) + (-4x^2 + 7x - 6)$
   $= 9x^3 - 4x^2 + (-x + 7x) + (5 - 6)$
   $= 9x^3 - 4x^2 + 6x - 1$

89. $2x^2 + x - 9 = 2(3)^2 + 3 - 9$
                $= 2(9) + 3 - 9$
                $= 18 + 3 - 9$
                $= 12$

90. $y^{16} \cdot y^{21} = y^{16+21}$
           $= y^{37}$

91. $(x^{16})^4 = x^{16 \cdot 4}$
       $= x^{64}$

92. $(b^7)^3(b^3)^7 = b^{7 \cdot 3} \cdot b^{3 \cdot 7}$
             $= b^{21} \cdot b^{21}$
             $= b^{21+21}$
             $= b^{42}$

93. $8a(a^2 + 4a - 10)$
$= 8a \cdot a^2 + 8a \cdot 4a + 8a(-10)$
$= 8a^3 + 32a^2 - 80a$

94. $(x + 8)(x + 3) = x(x + 3) + 8(x + 3)$
$= x \cdot x + x \cdot 3 + 8 \cdot x + 8 \cdot 3$
$= x^2 + 3x + 8x + 24$
$= x^2 + 11x + 24$

95. $(x - 7)(x + 7) = x(x + 7) - 7(x + 7)$
$= x \cdot x + x \cdot 7 - 7 \cdot x - 7 \cdot 7$
$= x^2 + 7x - 7x - 49$
$= x^2 - 49$

96. $(3y - 1)^2 = (3y - 1)(3y - 1)$
$= 3y(3y - 1) - 1(3y - 1)$
$= 3y \cdot 3y + 3y(-1) - 1(3y) - 1(-1)$
$= 9y^2 - 3y - 3y + 1$
$= 9y^2 - 6y + 1$

97. $(b + 10)(b^2 - 10b + 100)$
$= b(b^2 - 10b + 100) + 10(b^2 - 10b + 100)$
$= b \cdot b^2 + b(-10b) + b \cdot 100 + 10 \cdot b^2$
$+ 10(-10b) + 10 \cdot 100$
$= b^3 - 10b^2 + 100b + 10b^2 - 100b + 1000$
$= b^3 + 1000$

98.   The GCF of 21, 49, and 98 is 7.
The GCF of y^5, y^4, and y^{10} is y^4.

The GCF of $21y^5$, $49y^4$, and $98y^{10}$ is $7y^4$.

99. $15x^6 - 45x^2 = 15x^2 \cdot x^4 - 15x^2 \cdot 3$
$= 15x^2(x^4 - 3)$

100. $10a^4 - 30a^3 - 20a^2$
$= 10a^2 \cdot a^2 - 10a^2 \cdot 3a - 10a^2 \cdot 2$
$= 10a^2(a^2 - 3a - 2)$